MOTHER

of

GOD

MOTHER

of

GOD

AN EXTRAORDINARY JOURNEY INTO THE UNCHARTED TRIBUTARIES *of the* WESTERN AMAZON

PAUL ROSOLIE

HARPER

www.harpercollins.com

HarperCollins books may be purchased for educational, business, or sales promotional use. For information, please e-mail the Special Markets Department at SPsales@harpercollins.com.

FIRST EDITION

Designed by Michael C. Correy

Library of Congress Cataloging-in-Publication Data has been applied for.

ISBN: 978-0-06-225951-6

14 15 16 17 18   OV/RRD   10 9 8 7 6 5 4 3 2 1

To my parents, Ed and Lenore—
you have given me everything.

CONTENTS

INTRODUCTION

In January 2006, just eighteen years old, restless, and hungry for adventure, I fled New York and traveled to the west Amazon. With a satellite phone that my worried parents had insisted on renting, and a camera I had borrowed from a friend, I was indistinguishable from any of the scientists and tourists aboard that first flight. In all likelihood, like them, I would spend my three weeks in the jungle, and then return forever to my life back home. There was no way to know I was beginning a journey that would span a decade and take me to some of the most inaccessible reaches of the Amazon River at the most crucial moment in its history.

It has been a journey filled with unfathomable beauty and brutality that sounds more like fiction than fact: lost tribes, floating forests, murdering bandit-loggers killed by arrows, insectivorous slashing giants, and a secret Eden. There would be fistfights, stickups, and beheadings; new species discovered, fossils unearthed, and people riding on giant snakes. I would see places that no one had seen before, and cultivate a unique relationship with the secret things of the Amazonian wild.

I wrote this book careful to avoid it becoming a scientific text, or historical summary of the Amazon—other authors have written such volumes far better than I ever could.

Instead, I chose to focus on the extremes of adventure, and the beauty of wildlife and natural systems. The events in the pages ahead are written as I experienced them. Aside from changing a few names, dates, and geographic details to protect people and places, everything that follows is true.

BOOK ONE
THE AGE OF INNOCENCE

1
The Jaguar

*The few remaining unknown places of the world exact a price
for their secrets.*

—COLONEL PERCY FAWCETT

Before he died, Santiago Durand told me a secret. It was
late at night in a palm-thatched hut on the bank of the
Tambopata River, deep in the southwestern corner of the Amazon Basin. Beside a mud oven, two wild boar heads sizzled
in a cradle of embers, their protruding tusks curling in static
agony as they cooked. The smell of burning cecropia wood
and singed flesh filled the air. Woven baskets containing monkey skulls hung from the rafters, where stars peeked through
gaps in the thatching. A pair of chickens huddled in the corner,
conversing softly. We sat facing each other on sturdy benches,
across a table hewn from a single cross section of some massive tree, now nearly consumed by termites. The songs of a
million insects and frogs filled the night. Santiago's cigarette
trembled in his aged fingers as he leaned close over the candlelight to describe a place hidden in the jungle.

He said it was a place where humans had never been. Between rivers and isolated by a quirk of geography, it had remained forgotten through the centuries. The only tribes who knew of the land had regarded it as sacred and never entered, and so it had remained untouched for millennia. Decades earlier, after weeks of travel up some nameless tributary, Santiago had come to its border. There, he said, you could watch jaguars sunning themselves on open beaches in the morning; harpy eagles haunted the canopy and flocks of macaws filled the sky like flying rainbows. The river was so thick with fish that you could scoop up dinner with your bare hands. What he described was a lost world. He told me that it was the wildest place left on earth.

Don Santiago, as I knew him, even at the age of eighty-seven, would often spend months out in the jungle alone. He knew the medicinal properties of every herb, orchid, and sap in the jungle that surrounded the small indigenous community where he lived, in the lowlands of southeast Peru. He possessed an insight to the secrets of the forest greater than anyone I have met. He had lived in the jungle before boat motors or chain saws were available, before Spanish extinguished the native dialect of his people. Over the course of a long life in Amazonia he'd seen tribes that most people didn't know existed and species yet to be described to science. As he spoke of the jungle's secrets, lore of an age nearly ended, candlelight reflected from within shrouded sockets, the map of tributaries in his weathered face as cryptic as the landscape in which it was wrought.

I knew from experience that Don Santiago was never wrong, and in the years to come what he said had a profound effect on my life. As a naturalist, I knew that finding and sampling a truly isolated area of rainforest could redefine the baselines scientists use to study wildlife, and help me to protect habitat. As an explorer, it was the ultimate mission, and planning an expedition to find the lost world Don Santiago described became an obsession for years to come. But what kept me awake at night was something deeper than academic discovery or adventure. It was the realization that we could be the last generation to live in a world where such places exist.

I knew that it was a journey I had to make, and I knew I had to make it alone. If the place Santiago had described really did exist, as pristine and hidden as he had said, there was no chance I was going to foul the silence with the pollution and din of an entire expedition team: motors, voices, fuel. For a long time I struggled to work up the nerve. Even after having worked years in Amazonia when living in the bush had become second nature, the thought of going it alone made me shiver. There were too many stories, too many hundreds of would-be explorers, lost tourists, and even locals, who were swallowed by the jungle each year never to be seen again. In the most savage and dizzyingly vast wilderness on earth, the rule is simple: *never* go out alone. Yet there are those among us who have difficulty accepting what we have not found out for ourselves, who pass a WET PAINT sign and cannot help touching the wall. We simply have to *know*.

Only months after Don Santiago told me his secret, there I was: a hundred miles from the most remote human outpost, in utterly untouched, untrailed jungle. I was completely lost and terrified. I looked up hoping to see blue, but the entire sky had been eclipsed for days. From beneath 150 feet of canopy, the view above was a churning mess of understory vegetation, vines, bromeliads, and towering pillars of ancient trees swaying menacingly in the hot wind. In the Amazon less than 5 percent of sunlight reaches the forest floor on a clear day, which this was not. Black storm clouds lay pregnant across the canopy so low that vapor and branches intermingled.

I walked fast, machete in one hand and compass in the other, praying to glimpse a gap in the foliage that would signal the salvation of the open river. Earth, forest, and sky were all fiercely animated and moving in concert. Twigs and leaves, Brazil nuts, and even small animals rained from above. Trees as thick as school buses buckled and groaned, shaking the earth as the wind tore at their branches. I felt trapped. I longed to see open space. I'd been lost for days.

The storm was gathering force and my heart was pounding. One hundred feet to my right a branch the size of a mature oak snapped and hit the earth with the force of a car crash. More than once a cannon blast sounded as an entire tree split and fell. In the Amazon large trees are meant to topple, opening gaps for light, which allows new vegetation to flourish, while the carcasses of the fallen giants are digested by legions of insects, fungi, and proteins. It's how the jungle works; it's a giant meat grinder. When you are in it, you're part of that system, part of the food chain.

If the storm intensified, there was little chance I'd survive the resulting carpet bombing of shed tree limbs. Some of the great explorers have claimed that snakes or piranhas or jaguars present the gravest threat in the Amazon, but these declarations betray inexperience. The trees themselves, in their dizzying innumerability, isolate and disorient you, and in a storm prove the most deadly. Some of the true giants are so interlaced with vines and strangler tentacles that when they fall, their weight tears down almost an acre of jungle. There is no way to escape.

Please let me find the river, I whispered through clenched teeth as I hiked on. My route, had things gone as planned, ought to have led me out of the jungle and onto the open tributary yesterday morning. But as the weak light began to fade, it grew clearer that I would be spending another night lost in the Amazon.

Even as I employed every ounce of my consciousness and skill to finding some sign of the river, of open space and salvation, I could see the headline: "Twenty-One-Year-Old New Yorker Vanishes in the Amazon." Nothing would be worse than being picked apart by the armchair analyzers who surely would categorize me as just another yahoo kid who went out into the *wild* to find "nothing but mosquitoes and a lonely death." Or the way they criticized the guy who lived with wild grizzlies until being eaten. There is a difference between knowing what you are getting into and doing it well, and just flying out there and obligating others to clean up the mess. There are simply too many of them: people who court disaster in the name of adventure, getting themselves into trouble and then calling for help, or dying. No, I'd be quite happy to deny Jon Krakauer or Werner Herzog another project. True, I was young and the risk was moronically high; but this was something I had to do.

There were blisters on my hands from twelve hours of hiking and slashing, and my backpack straps had worn through the shirt and skin on my shoulders, but I pushed on a for a bit longer. I threw a handful of nuts into my mouth but was too dehydrated to chew and spat them out, even though I needed to eat. After twenty-three days in the jungle, three of them lost and alone, I had lost more than pound per day. In the last few hours I was certain I'd lost v more.

I needed water, but because of the rains, every water source was a turgid mess of sediment and detritus. Even small streams that should normally be clear were roiled and murky. For hours I had kept an eye out for the species of bamboo that fills its segments with water. You can spot it easily, leaning over from its own weight. When I found a patch of bamboo that looked right, I cut the stalk of one and water burst out. Hefting the pole I felt that all its segments were heavy with water, like a dozen tallboy beer cans stacked atop one another. I cut segments one at a time, guzzling down the contents. Just two bamboo poles supplied me with a belly full of water and enough to fill my bottle. Then I pushed on. It was getting dark.

After another twenty minutes of desperate slashing and hiking, another surge of panic and rage came as the realization broke that I wasn't getting out tonight. I slung my hammock, removed my shoes, and used a shirt to towel off my soaked body. It was impossible to tell if the storm would turn on full blast or if it would continue to simmer and growl the whole night. The image of my hammock being smashed into the earth by a falling branch played on loop in my mind. There was nothing I could do about it. I needed sleep.

Inside my hammock I zipped the mosquito net and spent several minutes killing the bloodsuckers that had made it inside. I went over a mental checklist. My machete was beside me on the ground, my headlamp was on my head, my backpack and shoes were hung off the ground to reduce the likelihood of them being shredded by ants. I opened my journal out of habit but abruptly closed it, too ashamed to admit to the page how miserably scared I felt. In entries from previous days, only forty-eight hours before, I was living my dream, on a mission, soaking in every sensation of being immersed in the gut of the jungle. But confronted with the cosmic force of the coming storm and the reality of being truly lost, my courage stores were waning.

Anyone who has seen, or even read about, what the Amazon is capable of during the rainy months would know that attempting even the most mundane travel is virtually pointless. Cities and towns flood, dirt roads become muddy rivers, and actual rivers can swell more than fifty feet in places, exploding far onto land. Larger tributaries can burst their banks and flood miles of forest, ripping thousands of trees from the earth in the all-encompassing current. The result is a river of giant timber that would turn a boat to splinters. Before starting the expedition, the one now veering dangerously off course, I had known these dangers but saw no other option. Time was running out.

For months the sound of heavy machinery and chain saws had grown louder; smoke could be seen on the horizon. After thirty years of dormancy the trans-Amazon highway was under renewed construction; the final link was being constructed over the Madre de Dios River. For the first time in history the heart of the Amazon would be connected by a land trade route to the Asian market. Offshoots of the highway were rapidly metastasizing throughout the lowlands as colonists cut their way into the frontier. Towns were filled with indigenous protesters, police in riot gear, and people were dying. Don Santiago would soon be gone and it seemed that an age had ended. The western Amazon was under siege.

With a light pack made up of ten days' worth of food, matches, machete, bowl, camera, and hammock, I had hitchhiked as far into the jungle as poachers would take me, and then plunged into the trackless green. Maybe I'd been born a century too late? What if my destiny was not to protect the west Amazon but to bear witness to its annihilation? Tucked into my journal was a hand-drawn map with a circle drawn at the place I had come to call the Western Gate, the boundary of the nameless Eden. I was twenty-one years old, young enough and dumb enough to voluntarily trek into the Amazon, old enough to know that what I sought was worth the risk.

I lay in my hammock looking up as darkness consumed the jungle. Lightning in the low clouds flashed emerald green through the leaves, turning the canopy to a ceiling of stained glass. As downdrafts gusted hot and then cold air through the subcanopy, the light show made the savage landscape all the more surreal. I closed my eyes and told myself that everything would be fine. After four years of living and working in the jungle under the tutelage of the Ese-Eja Indians, I knew what I needed to survive. But even so I could hear my mother's voice from years earlier warning that even the best swimmers can drown.

I don't know how or when, but eventually long, torturous hours of blackness morphed into unconsciousness. For a time there was peacefully nothing. But then, prompted by some dread instinct, I awoke to a nightmare. My eyes were open, but nothing was visible in the inky void. For a moment I wasn't sure where I was or why I wasn't sleeping. I wanted to call out, "Where the hell am I?" Then, as my mind slowly booted up, I remembered. Oh yeah, in the Amazon, alone . . . except I wasn't alone.

The Madre de Dios, or Mother of God, is a living anachronism. Like a world made from Joseph Conrad's nightmares, it is the edge of nowhere, a vast region choked in snarling ancient jungle. Nestled in southern Peru under the shadow of the Andes to the west, with Bolivia to the south and the Brazilian state of Acre to the east, it is remote, pristine, and like nowhere else on earth.

Some say the southeasternmost region of Peru got its name because an apparition of the Virgin Mary appeared to a Spanish conquistador in the late 1500s. Others maintain that the isolated no-man's-land was simply given a "God's country" designation for being wild and unexplored. Still others say that the name was given out of reverence, that even the conquering Spanish were overwhelmed by the raw wilderness and unfathomable bounty of the jungle there. One thing is certain: in today's context the profound name remains worthy, for the region is the womb of the Amazon.

To properly appreciate the scope of topographic magnificence of the Madre de Dios you'd need to imagine cramming the varied temperature range contained within the latitudes between Peru and Alaska into a dozen miles: from frozen peaks to steaming jungle. In the western Amazon, glaciers in the high Andes send mineral-rich runoff in torrents toward the land below. These streams and rivers rush through mossy cloud forests and down into the flat lowlands, where they converge to create the Madre de Dios River and begin the slow march 1,400 miles across the continent, bursting into the Amazon's main channel roughly twenty miles downstream of Manaus, Brazil.

The tropical Andes and the lowland Amazon are considered two separate, mega-biodiverse biomes; entirely different ecosystems. It is the intermingling of these two systems in a tropical climate, with abundant moisture and in drastically varying elevation, that makes the perfect storm for speciation. On a clear day from the Los Amigos River, a tributary of the Madre de Dios, it is possible to look west over the boiling lowland jungle and see the snowcapped Andes looming divinely far in the distance. Contained in that single view is the greatest array of living organisms to have ever existed.

Amid the foliage of the Andes/Amazon interface, which constitutes more than 15 percent of the global variety of plants, is a land of faunal giants. In the canopy harpy eagles hunt for sloth and red howler monkeys, the latter the size of small children that the eagles skewer and lift into the air en route to be dismembered in the nest. Toucans greet the mornings, and stunning blue-and-yellow and scarlet macaws are like flames in the sky. Each rainy season frogs descend from the canopy to breed in stagnant forest pools, and in the dry season butterflies flock in clouds of thousands on the riverbanks in color variations that would stun a rainbow.

Most ecosystems have a single, indisputable apex predator, but the western Amazon is more like a cage fight in murderers' row. With so much muscle around, they've had to split up the terrain. The harpy eagle takes the canopy, while jaguars cover the ground. Anacondas and black caiman crocodiles, which can reach eighteen feet in length, battle in the rivers and lakes, which are also haunted by giant otters, a formidable hunter whose Spanish name translates to *river wolves*. Probing the deepest parts of the river are 150-pound black catfish. And yet this list of killers is far from complete: several other species of cat, croc, mammal, and large snake back up the hulking lead characters. The list of smaller hunters is virtually infinite.

The rough tallies for the entire Andes/Amazon region: 1,666 birds, 414 mammals, 479 reptiles, 834 amphibians, and a large portion of the Amazon's 9,000 fish species. In the Madre de Dios alone there are more than 1,400 butterfly species. The numbers on everything from bats to beetles are constantly changing as scientists learn more. In early 2012 researchers from Conservation International announced that they had documented 365 previously unrecorded species in a single study area, just a pinprick of the landscape.

Within the impenetrable assemblage of giant hardwoods and bamboo are palms with ten-inch thorns that can run a man through, and others that walk from place to place beneath the canopy on their roots, like Tolkien's Ents. Flowing in the cambium of some trees are poisons that can kill you in minutes and other compounds that can drop your mind into hallucinogenic pandemonium, but there are also medicines that can control fertility and cure the most horrendous diseases. The west Amazon was where the first cure for malaria was discovered and where rubber was first tapped on a large scale. Amid the diversity grows one tree that produces sap almost completely made of hydrocarbons, producing 1,500 gallons of sap each year that can be poured directly into a diesel motor as fuel.

While the food chain can be mapped in web format for most ecosystems, the west Amazon defies human explanation. If it were possible to trace the elaborate interspecies relationships unfolding within the jungles there the result would most likely resemble a Jackson Pollock painting the size of Rhode Island. Even today we know little about the system. It is for this reason the region has been described as the "largest terrestrial battlefield" on earth. Every last organism is eaten. Life here is a countdown, a temporary stasis as the jungle waits, inevitably adding all things into the rapid cycle.

The Mother of God is a region of extremes, polarizing all elements within it, including humanity. In a world where rivers are highways, there remain infrequent indigenous settlements. It is not uncommon to come across a cluster of palm-thatched villages among the green. In the early morning you can see women washing clothes in the river, while children play and splash. Men hunt and fish and grow crops like bananas and yucca, or collect Brazil nuts. Much of the sparsely inhabited backcountry is like this, peaceful, and made up of simple, warm, friendly people.

At the other end of the spectrum are the extractors. Prowling the backcountry is a host of loggers, drug traffickers, poachers, and gold miners. The latter use motor pumps to tear up the riverbed sand in search of their prize, dumping mercury into the water and polluting the otherwise pristine world. *Narcos* also make strategic use of the geography. It is rumored that the cocaine trail comes up from Bolivia and that the runners use a small airstrip hidden in the jungle. They apparently maneuver the plane into a gap in the canopy and land on a runway obscured by the branches above, remaining invisible to aerial surveillance. But it is the loggers who are the most blatantly nefarious. There are numerous accounts of loggers clashing with local people and even isolated tribes, native arrows little match for loggers' modern guns.

The collision of human worlds is comparable to the westward expansion of European settlers across North America in its components, but the situation in Madre de Dios is wilder by several degrees of magnitude. Copy and paste the players from the American West into the insane context of the Amazon, change a few names, sprinkle in some anacondas and several million other species, and the similarities are eerie. Loggers versus Indians, gold miners versus helicopter commandoes; oil companies, pipelines, new roads, secret genocide, corruption, greed, missionaries, bandits, politicians, and massive paradigm shifts unfolding at a dizzying pace.

At a time in history when scientists are recording unprecedented extinction rates and many people feel that the loss of biological diversity and deteriorating natural systems is the defining issue of our time, the west Amazon is ground zero. Nowhere are the stakes higher.

It was within the depths of this world that I slept on that torturous night. All hope of finding Santiago's wild land had faded. Now the great adventure had become a survival situation, a question of direction and luck. I remember waking into darkness like the belly of a black hole. Alone in my hammock, I listened desperately. Something was nearby, something big. I could hear breathing. I shut my eyes. Heartbeats shook my chest, and my blood rushed audibly. I had no thoughts, only blind terror. The volume of air drawn with each sniff told me this was something massive. My nostrils filled with a pungent odor as my hand instinctively went toward my headlamp, making a small noise against the hammock's fabric.

A growl erupted from the darkness. A god's voice. Warm breath fell on my neck in savage staccato like thunder, cosmic and overwhelming. Every fiber of my body understood the command of that growl: don't move. I closed my eyes and lay still, too terrified to move. Cradled in blind purgatory, grasping at lucidity, I was helpless and prayed that whatever happened next would be over quickly.

In the context of their rainforest environment, jaguars are ghosts. Masters of the shadows, they employ a skill for stealth that is leagues beyond human ability. Scientists who study jaguars their entire lives can go years without a sighting and instead have to rely on tracks and scat, and on camera traps and radio collars to collect data. The cats are thick-bodied and powerful, the pit bulls of the big cats. They move silently over land and through water. They can drag a deer up a tree and, at 250 pounds, can overpower anything in their environment. To a jaguar, dispatching a human wrapped in his hammock would translate roughly to you or me peeling a banana.

I could feel her breath as she drew my scent into the labyrinthine reaches of her nasal cavity, her face only inches from my right ear. Is this how it ends? She sniffed and drew nearer, exhaling another furnace draft onto my neck. For a small eternity she was silent, standing invisibly beside me, incredibly close.

Lost and alone beneath the storm and the canopy, on the dark side of the planet, it was a pivotal moment of a story that began when I was very young and would shape my entire future. Despite the jaguar at my side, a remarkable calm ebbed through me. She hadn't come for blood.

2

Restless

Earth and sky, woods and fields, lakes and rivers, the mountain and the sea, are excellent schoolmasters, and teach some of us more than we can ever learn from books.

—JOHN LUBBOCK

When you travel east from Lima by air, floating over the snowcapped steeples of the Andes, the clouds are numerous and blinding. Jagged parapets draped in glaciers fall to immense valleys that seem to yawn into eternity. The land between the great peaks is stoic and barren, treeless and empty for thousands of miles at a time; an alien landscape interrupted only rarely by a long and lonesome dirt road.

The FASTEN SEAT BELTS sign came on as the pilot threaded between mountains that seemed too massive to be real. I was vaguely aware that the guy in the seat next to mine was pale green and burping bile from the turbulence. Behind me a woman was praying the rosary in Spanish. The plane's wings were flapping as we were jerked up and down. But I was mesmerized and barely noticed. I had waited my whole life for this moment.

The empty valleys below were gradually shading green. The barren emptiness began to turn to lush foliage and then riotous cloud forest as we lost altitude and the mountains dropped off. Vision came in glimpses through the clouds as glacial rivers cascaded through a world of moss and mist. Then came more clouds, torturous moments of whiteout when I could feel my heart pounding with anticipation. Then it happened.

Shuddering through turbulence, the plane dropped below the ceiling, revealing an unbroken immensity of green jungle from horizon to horizon. For the first time in my life the breath was sucked from my chest. Rivers lay across the range in great sweeping arcs like bronze serpents reflecting light skyward. Mist sat like puddles scattered in the omnipresent foliage, rising and curling in places. It was like looking into the vault of the universe to where all the greatest secrets were kept, the library of life.

How could there be so much jungle? Since childhood I had dreamed of this, wondering what the Amazon would look like, smell like, and feel like; imagining, hoping, waiting but never once gaining any preparation for the mind-blowing reality of setting eyes on it for the first time.

People always ask me how I came to work in the Amazon at such a young age. It is a difficult question to answer, because the trajectory that sent me into the jungle started when I was very young, and had something to do with a dismal cloudy day in high school, when a teacher threw me up against a cinder block wall. No one else was around, just him and me. His hands were around my neck and for a moment I paused in disbelief, watching his ugly gray eyes, so revoltingly close to mine, bulge in anger. In that moment I was terrified, but not for me. I was scared because I could feel the cerebral rage dispatching through my limbs, and the imminent reaction that I knew would be out of my control.

But I hadn't always been violent. I started out as a gentle, nature-loving dyslexic kid from New Jersey. Actually, I was physically born in Manhattan, back when my parents were living in Brooklyn, and I have always identified the latter as where I'm from. Even after we moved to Wyckoff, a town in the New Jersey suburbs, we'd still make it to Sunday dinner every other week at Grandma's house in Brooklyn, under the shadow of the Verrazano-Narrows Bridge, with all the aunts, uncles, and cousins, for the best food in the universe.

Both of my parents were teachers, but my mother became a full-time mom after my sister and I came into the picture. My dad kept his job in Brooklyn and made the hour-and-a-half commute each way for almost twenty years. When he got home from work each day my mom would be in the kitchen, filling the entire house with the scents of garlic and marinara. We'd all play and laugh and eat together, every night. My dad and I were always wrestling and playing ball. We built snow-men together, hiked, and sometimes he'd make us all laugh until it hurt by quoting dramatic passages by Dostoyevsky.

Though I wasn't diagnosed until high school, I am dyslexic, which disability caused me to struggle throughout my school life. It took me far longer than everyone else to learn to read, and math was like Sanskrit. School was always a nightmare for me. I just wanted to be outside, and as a kid I'd look out the window until the teacher yelled at me. So then I'd draw, which also wouldn't end well. That's how I ended up doing first grade twice, in three different schools.

While I bounced around the education system, my parents took teaching into their own hands. I remember weekend hikes with my parents and sister, along with our golden retriever, Sam. I was always in search of bugs, frogs, salamanders, and snakes. For me, exploring the forest, overturning rocks and stones, and being in the woods was freedom. As I grew, my delight in the natural world intensified. I continued to develop a keen eye for things other people didn't see, and a gentle touch needed to handle small forest life.

My parents encouraged my love of the natural world, and I can remember my mom spending hours explaining the difference between African and Asian elephants, black bears and grizzlies, and other basics to a captivated five-year-old me. She used to blindfold me and have me identify the trees in our backyard by the texture of their bark. The big old oak that leaned over our house, the maple that my sister and I liked to climb. I was good at that.

I cried when she read me an article about Lonesome George, a Pinto Island Galapagos tortoise. The article explained that he was the last of his kind and that when he died his species would be extinct. I was well under ten years old, and the thought kept me awake for weeks. That a species could be removed from existence by humans, or that an ecosystem could vanish before I had the chance to see it, horrified me. I grew up with a sense of urgency. I wanted to see the world's wild places and creatures before they disappeared. Thankfully, at that age there were many things I didn't know.

While still in grade school I'd spend months each summer caring for and raising praying mantises, the *T. rex* of the insect world. Threatened by pesticides and habitat loss, they were the first endangered carnivores I worked to protect. Each spring

I'd hatch mantis nymphs and release several hundred of them into the wild and keep twenty. Over the summer I'd feed them insects, and they'd brutally cannibalize each other; by the fall I'd be left with one or two giant, brilliant green, four-inch-long carnivores that were eating butterflies, katydids, and steak. One time after they had been mating for days, I watched a female rip a male mantis in half, and then eat him. I drew and studied my mantises and learned everything about them. I bred them, wanting to help boost their numbers in the wild. I spent time rehabilitating injured or orphaned animals. One time in my early teens I found a six-foot-long black rat snake (*Pantherophis alleghaniensis*) that had been attacked by a dog and gravely wounded, and kept him on a shelf in my room for two months while he convalesced.

Even as a child I considered myself the keeper of things meek and wild. On summer vacations buying two eels from a bait shop and releasing them in the ocean became a tradition. Hopping out of the car to help turtles cross roads before they were run over was a must, as was, later on, secretly freeing birds and monkeys from the snares of poachers. Throughout my life I would too frequently find myself holding tiny lives in my hands. It was as if they were drawn to me, and I to them. Regardless that the creature would most often have no concept of my help, in whatever crossroads of fate and luck that exist in this world, I felt a deep responsibility to protect. Most often, of course, it was my own species that was the reason for the imminent tragedy to the small life that so desperately needed an ally.

Once I encountered a fisherman unloading his catch, and beside him was a magnificent guitarfish, a graceful kite-shaped type of ray with a spike-ridged back, gasping in the sand. The ray was of no use to him, but instead of returning it to the ocean, he had been callous or lazy enough to simply and needlessly let it die. I spent a half hour ducking and gasping beneath the breakers while holding the beautiful alien in my arms, oxygenating its gills until it regained the strength to swim. *You'll be all right*, I whispered. *There is at least one human who is not a savage. You'll be all right.* And at last it was.

The older I grew, the more I was drawn to nature. I was mystified by how sharp my own senses could become beneath the towering tulips, oaks, and sycamores of New York and New Jersey. As a kid I spent an inordinate amount of time chewing on big questions, wondering, exploring. Watching sunlight refract, or a doe nurse her fawn; I was deeply fascinated, illuminated, by the world I saw around me. Though my mom dragged me to church every so often, the woods were where I felt close to whatever energy pulses through life. Turn-of-the-century naturalist John Burroughs wrote, "We now use the word Nature very much as our fathers used the word God," a sentiment that developed organically in young me.

Somewhere from the childhood haze that is both vivid and fleeting emerges one day that I remember in almost perfect detail: visiting the Bronx Zoo for the first time—specifically, the Jungle World exhibit. I had dragged my parents around for hours, talking nonstop, fascinated by everything I saw. But Jungle World shut me up. I couldn't have been more than eight years old but I can still feel my own spooked awe as I walked into the rainforest house through dark curtains, hearing the calls of hidden primates, thunder, and insects. There were giant kapok trees and hanging vines; brilliant arrow frogs and other creatures I'd never heard of, all presented in a world crafted by experts to mirror jungle habitat. It was pure magic.

How is it that places and creatures we have never seen can resound in our innermost depths?

I remember struggling to imagine that such a place could truly exist. The world as I knew it was pedestrian, manicured, and predictable, not wild and mysterious. But my skepticism dissipated in the face of abundant proof: photos of scientists working the field, dirty and dedicated in distant countries. In particular, the photograph of a half-dozen men holding a twenty-foot-plus reticulated python burned onto my mind. For those people life was an adventurous and purposeful quest, out in the jungles at the ends of the earth, saving species. Moving through the shadows from one exhibit to the next, I experienced something I had never felt before: belonging. It was perplexing at my young age to experience such powerful gravity toward a world I had never seen and barely believed existed. There were no words, only an innate recognition of coded bearing, as powerful as the instinct that guides a hatchling sea turtle into the throbbing counsel of the poles. It was a unique moment of orientation in an obtuse childhood, which I held on to like a treasure, through many dark winter months sitting at a desk, and over many years.

As I got older my ambition began to boil and my fight with the education system intensified. I wasn't the only one who suffered. I would look around the classroom—in the dismal stillness of a teacher's droned lecture, or worse, as the scratch of pencils made the only sound—to see dozens of kids in slow atrophy. Of course, there were those who didn't mind it, or were too young to question the system, as well as those who seemed to actually thrive in the structure—watching them made me feel all the more dysfunctional. What was it they had that I lacked? However, I wasn't alone in my struggles, and over the years saw many brilliant and creative young minds bound by walls, rules, conformity, and endless boredom. Artists, musicians, athletes, farmers, and free spirits of every kind have been hammered into submission by an archaic, outdated system. Even today it kills me to watch kids drag through school during the years when they should be out in the world, experiencing and learning. What does education do? As Thoreau famously answered: "It makes a straight-cut ditch out of a free, meandering brook."

Thankfully, though, there was relief. In my early teens I'd spend entire weekends in the forests and rolling hills of Ramapo Reservation and Harriman State Park with my friend Noel. We'd been friends since my third stint in first grade and by age thirteen were heading out into the woods with nothing more than hunting knives, a few steaks, sleeping bags, and my dog, with the intention of getting as lost as possible. Those adventures sustained me. I needed adventure: not vacation, not distraction, but true, meaningful adventure. We had some good ones. After getting lost in the woods together, we'd build a place to sleep out of sticks, sit up by the fire at nights, and scramble over mountains sometimes for days to find our way out. Through the glowing green summers, the oranges and yellows of autumn, and through the bare frozen mountains in the lonesome winters we forged our navigational skills, resourcefulness, grit, and a powerful friendship.

Through middle school and freshman year of high school I broke all kinds of records for detentions and suspensions and made my way to each June feeling I had barely survived. By tenth grade I couldn't do it anymore. To keep my brain from atrophying I read books during class. I took refuge in historical figures who had also felt trapped or alienated by early education, especially the ones who were passionate naturalists like Teddy Roosevelt, Albert Einstein, and Ben Franklin. I read about people like Winston Churchill, Varian Fry, Jane Goodall, Alan Rabinowitz, and Steve Irwin; people who had sought adventure and purpose in life and had really *lived*. In the Shire of my world people went to the grocery store, discussed cell phone plans, golfed, and watched sitcoms. It was comfortable, clean, organized, and safe. Though I understood the privilege it was to live in a stable place so far away from the world's troubles, I wanted out. I didn't want to be safe; I wanted to have the shit scared out of me.

Around sophomore year of high school my grades deteriorated to the F range. I was failing. Things with my parents, whom I'd always been so close with, started to fall apart. They knew me well enough to know that if I failed, there was no way I was going to summer school, and then I'd just be a dropout. I was furious, depressed, and losing the only anchor I had ever had. The dinnertime laughter and warmth of childhood had been replaced by fights that rocked the walls each night. I remember wondering if things would ever go back to the way they were.

Being thrown against the wall by my science teacher was the end result of a yearlong battle of wills. I had been reading inconspicuously in the back of the room. Dr. Sherk had been droning on for the second period in a row when he stopped his lecture, walked toward me, and grabbed the book from my hands. He returned to the front of the room and into an office space to stow the book. I followed. Inside the office, with the door closed, he held it behind him like a child playing keep-away. "I wasn't disturbing the class, or affecting you in any way; give it back," I said. He pointed his finger at me.

"Get out of my office."

"Give it back."

"Get out now or I'm calling security."

"Fuck you, just give me the book!" I never expected all five feet and two inches of him to come King Kong across the room and grab me by the neck, but he did. I launched back and sent him across the room to the opposite wall, where I watched his face turn colors as I choked him. I wanted to bury my fist in his face. I was furious but on some level liked the feeling of excitement. What would happen if I did it? What if I just smashed his face with everything I had? Surely the cops would be involved, maybe I'd be expelled, but surely the monotony would be broken for once. That was all that mattered. I almost did it. But instead I released him and let him fall coughing onto the floor. Later he was fired for throwing a metal scale at some kid, I heard. But the incident scared me. I was losing my patience with life.

Not long after, I was wandering around the hallways and stopped by to see friends in shop class. There were wood and tools on every surface and sawdust on the floor. It smelled like my basement at home, where my dad did his woodwork. It was then that an offhand joke made by one of the nerdy-looking juniors halted me in my tracks. One guy botched cutting a smooth heavy piece of wood, and the junior started laughing, saying, "It's a good thing we're mowing down the Amazon rainforest so that we can dick around with wood in class." Several of them laughed.

Those three syllables hit my ears like a flood: *Amazon*. I knelt and delicately lifted the piece of wood as though it were some hallowed treasure and inspected it. As ridiculous as my fascination for the block must have looked to anyone watching, imagining that it may have *actually* come from the *Amazon* hastened the beat of my heart. An image of mist-shrouded jungle choked in vines and endless rivers, and a kinetic tingle of a long-forgotten dream, surged over me. *The Amazon.*

For the first time in years I remembered the photo of the scientists at the Bronx Zoo, and the sensation of walking through Jungle World. I remembered the hikes on rainy days with Noel when the forests were so dark and green I could pretend they were jungle. It was a piece of me that had been pushed aside amid the turmoil of teenage life, and was suddenly called back.

Whatever riches it is possible to possess in life, having parents who are behind you and who understand whatever it is that makes *you* is among the most valuable assets a child can have. As my grades dropped below the point of no return, and my total suspensions for the year hit double digits, my parents suggested I drop out and go to college. "Why not just take your GED and go to college?" my mom asked.

When classes finished in June it was the last time I'd set foot in high school. I never went back. I didn't even tell the few friends I had. I just left. My mom and I would laugh together when at the start of what would have been my junior year, my high school was calling each day to report that I hadn't shown up. By that time I had taken my GED and enrolled part-time in college.

No longer confined to a desk for eight hours a day, I felt free. I started working as a lifeguard at a YMCA in my town. I saved every penny I could and spent my free time contacting scientists and conservation organizations and combing the Internet for anyone who might need a researcher. But most researchers weren't interested in an untrained high school dropout, and everything else was tourism. I continued to search. I wanted to find the most isolated and remote spot possible. So, while my first semester of college progressed, I sent emails for months without ever getting a reply, and then, finally, got *one* back.

It was from a British biologist named Emma, who ran a research station in southeast Peru. In the email she apologized for being out of touch for weeks and said she had been in the jungle. She had a crew of student researchers heading into the jungle after Christmas for several months and included detailed information about the research they'd be working on. I responded that it sounded perfect and that I had a month off between semesters. A few days later Emma replied to say that it wouldn't work. They couldn't make the two-day journey out of the jungle to bring me back to town, even after a month. Two-day journey? That was seriously deep jungle, and there was no way I could miss it. I spent a day agonizing over what to do and then lied and told her that I had already booked my airfare. She agreed that if I paid for gas, they'd make it work. I was in.

On a frigid day in late December 2005, I left my worried parents and the gray concrete world of JFK Airport. As that plane came low over the Madre de Dios River, I was a fish about to see water for the first time, a raindrop about to enter the river.

3
Into the Amazon

It's a dangerous business, Frodo, going out your door. You step onto the road, and if you don't keep your feet, there's no knowing where you might be swept off to.

—J. R. R. TOLKIEN, *THE LORD OF THE RINGS*

The door of the plane opened and the smell of jungle was instantly overpowering. It was the stench of a hundred billion trees, wet and rich. Walking across the tarmac somewhat dazed, I sucked in lungfuls of air magnificent in its floral richness.

Puerto Maldonado is a small city located in the middle of the Madre de Dios, nestled in a bend of the namesake river itself, as though being constricted by a titanic anaconda. Originally a gold mining outpost and then a center of rubber production, today it's the capital of the Madre de Dios. It is a dusty but quaint city that was connected by road to the outside world only in the 1960s. Even today many of its streets are unpaved and give the feel of a small jungle town. At night gypsies juggle knives at traffic lights, and women with large orange howler monkeys amble beside tourists, while on the fringes of town red lights illuminate young girls waiting for customers.

In the mornings bread carts travel up and down the main streets, followed by carts loaded with cages of small quails; their freshly laid eggs are hard-boiled and sold as snacks. The markets are bustling labyrinths constructed from blue tarp and corrugated steel. There the jungle's bounty enters the economy: gold miners, fishermen, peddlers of every known fruit, and hunters with fresh-killed bush meat converge there. Only a block from the Plaza de Armas you can look out over the Madre de Dios River and the jungle beyond.

The Puerto Maldonado airport is a single-story structure surrounded by foliage with two main rooms: arrivals and departures. The walls of the building aren't even walls: they are lattice, open to the air. As I claimed my bags on the small conveyor belt several tropical birds and butterflies moved around the room. I stepped out into the parking lot amid a gang of eager auto-rickshaw drivers and searched for my contact. She wasn't hard to spot. Tall, blond, and with cool gray-blue eyes, she was a head above most of the Peruvian men. "Yeah, nice to meet you!" she said in a melodic British accent, extending her palm.

Emma was born in Cambridge, but her family later moved to a small village north of London, called Bedford, where she grew up. After receiving her undergraduate degree in biology and earth science, and barely into her twenties, she set her sights on adventure and headed for South America, into the Madre de Dios. She had applied for and secured a resident naturalist position at one of the first and most celebrated research and tourism lodges in Peru, Explorer's Inn. The inn sits on the confluence of the Tambopata and La Torre Rivers, considered a gem of biodiversity even within the context of the west Amazon.

Upon arrival she had mere days to begin learning in the field from both scientists and locals, before starting work as a guide. Surrounded by an inspiring crew of like-minded people, all passionate about rainforest conservation, Emma found that she thrived in the jungle. After a few months she was speaking fluent Spanish, working on research projects, and leading tourists on adventures through the Amazonian wonderland. With her shrewd eye and trained ear she was able to expose creatures virtually invisible to the tourists she guided.

Once, after a long hike, Emma jumped into a lake for a swim. The doubtful tourists were hesitant to follow, but Emma waved them in. As they entered the Brit felt a rapid slash and a parting of skin on her foot; she stifled a shout. Once everyone was in the water playing and distracted she had a chance to inspect and found that a chunk of her toe had been bitten clean off, most likely by a piranha. She had to conceal the blood when putting her shoes back on.

In another instance of implacability, years after we first met, I was at Emma's house in Puerto, mending a hammock. She returned from the market bloody. She parked her bike and took off her helmet. I gaped and asked, "What happened to you?" She explained that a young local had tried to mug her, knocking her off her bike and into traffic. Her entire forearm and chin were raw. "Are you okay?" I asked, but she was already past me. From inside the house she called, "Paul, do you want some pasta, mate?" and would hear no more about the matter.

She was tough, pretty, and knew her stuff in the jungle, factors that made her irresistible to one of the handsome young indigenous guides who also worked on the Tambopata. Juan Julio Durand, or JJ, as I would later know him, was of Ese-Eja

Indian descent and had grown up in the jungle. His skill and innate awareness in the forest were borderline supernatural. Emma recognized that even among natives Juan had a special bond with the forest. He became the warm complement to Emma's cool demeanor, and in her first few months in Peru, the girl from Bedford, England, had found romance in the wilds of Amazonia.

Emma and Juan Julio learned and explored together. JJ had recently been discharged from mandatory service in the Peruvian army. He had seen action while away, including a police-versus-army shoot-out (welcome to Peru). In the exchange, he had been shot in the thigh, a wound he flaunted to Emma in the days when they worked together at various lodges up and down the Tambopata. But soon they hatched a dream to start their own lodge. At that time in Peru, land was cheap, and after some saving and scrounging they were ready. All that remained was finding the right place.

They launched expeditions all over the region in search of a location. The jungle is hardly a uniform thing, and even within the fantastic diversity of the west Amazon there are some areas that seem to glow exceptional. Surrounding Puerto Maldonado was only farmland. All of the truly wild areas required multi-day trips into nowhere. They had to find an in-between. Emma and JJ traveled days up the Madre de Dios to the border of Manu National Park, then south on the same current to the border of Bolivia, stopping to bushwhack through the jungle and search for signs of thriving primary forest. It was on the Las Piedras River that the forest spoke to them.

The Las Piedras is the longest river in the Madre de Dios, a three-hundred-mile squiggle through dense, unbroken jungle. It is so far off the radar that the only reference to the river in literature is from the early 1900s, in the book *Exploration Fawcett*, which describes one team of explorers that "crossed from the Tahuamanu to the Rio de Piedras, or Tabatinga. . . . In spite of the party's numbers, so many of them were killed with poisoned arrows that the rest abandoned the trip and retired. There is a tribe there called the Inaparis." As Emma and JJ struggled to start their lodge, the tribes Fawcett described were still active in the Piedras's headwaters, and still shooting arrows at intruders. But the river's remoteness and danger had kept it wild, and it was there that Emma and JJ had their breakthrough.

On their second day of forging by boat up the Las Piedras, they were rounding a bend when a screaming cloud of crimson suddenly engulfed them as seventy macaws took flight. Flashing cerulean tails, the birds launched into a long 360 before settling back onto the riverside cliff. Both Emma and JJ had worked with the endangered birds at Explorer's Inn and realized they had discovered something special: a *colpa*. Often referred to as clay licks, colpas are areas of exposed clay where birds and other animals come to feed on salt deposits. Species flock to these areas to replenish sodium in their bodies. Colpas attract herbivorous mammals, including monkeys, deer, peccary, birds, armadillos, and numerous other prey species for cats like ocelots, puma, and jaguars. These deposits are rare, and their locations are guarded secrets to the creatures and local people that know them. Colpas on the riverbanks can attract hundreds of macaws at a time, creating one of the most colorful natural spectacles on earth. Emma and JJ had just struck gold.

In days to come JJ pushed his powers to the limit as he and Emma explored the untrailed reaches of the land near the colpa. They found black spider monkeys, giant uncut hard-wood trees, massive herds of wild peccary, and copious jaguar tracks. They found *aguajales*, small forest swamps, as well as streams, and even a fifteen-foot waterfall. This was home. It was there that they began building their research station.

At that time the Peruvian government was eager to offer land concessions to make economic use of the vast unpopulated areas in its southeast region. The Brit-native team was able to secure first an ecotourism concession, then a Brazil nut concession, and finally a third parcel of land that brought the total area they protected to twenty-seven thousand acres.

Emma used her British contacts to bring schools and ecotourism groups to the new research station and base research projects there. Soon Emma and JJ were busy guiding travelers and Ph.D. candidates, protecting thousands of acres, and raising their son. Isolated from the world by hundreds of miles of jungle in every direction, brimming with wildlife, the Las Piedras Biodiversity Station, or LPBS, was the most beautiful place in the Madre de Dios.

On the two-day journey upstream in January 2006, I could barely keep my jaw shut I was so amazed. The jungle was endless. I had grown up in a world where civilization surrounded everything, and nature was confined to finite areas; but snaking through the jungle the inverse was true, to an immense degree. The reality of being a speck within thousands of miles of Amazonia was exhilarating. For eight hours the first day and six hours the second, we wound upriver through explosive green foliage. We covered miles of unbroken jungle, seeing people only briefly, when passing an indigenous village of thatched palm huts, where men in canoes hewn from solid logs waved to us.

There were caiman on virtually every beach, and dozens of turtles sunning themselves on logs with butterflies balancing on their noses. We saw monkeys in the treetops, kingfishers and herons alongshore, and every so often a pair of macaws would fly overhead. In a moment of hushed wonder we watched as a giant anteater paddled past our boat as it crossed the river, its periscope nose and long, bushy tail propelling it through the current like a bizarre seven-foot dragon. Even Emma couldn't hide her fascination; in more than ten years in the Amazon she'd only seen the massive animal twice.

On that boat were several other volunteers, students who had signed on to help Emma and JJ with their macaw research. I couldn't help asking Emma rapid-fire questions, until she slipped on her sunglasses and headphones and pretended to sleep, leaving me to gawk at the passing scenery. I was losing my mind in excitement; I wanted to stop the boat and run into the jungle. In the interest of making good time Emma would hear none of it. But when I spotted a large whip snake and called out, she was at full attention.

The snake was in some bamboo by the river's edge, and looked about nine feet long. Emma took one look and motioned to the driver to swing around and head for it. "JJ, get ready! Paul's going to teach you how to catch a snake!" Prior to coming I had told Emma that I knew my snakes and wanted to study them during the expedition, an offer she took seriously.

JJ had been dozing with eleven-month-old son Joseph on his chest. He sat up and handed Joseph to Emma. "Go to mama," he cooed while removing his blue alpaca wool sweater. "Okay we go?" he asked, and smiled at me. The boat pulled up next to the bank, and the large snake froze, waiting to see what would happen next. JJ took off his shirt and then pants, and slid into the water in his underwear. I followed, and though I hate to admit it now, I was feeling a little nervous, but not about the snake.

When you grow up in a place like New York, the heart of so-called civilization, the Amazon is burned into your mind as a place where some way or another, you are going to die. I tried not to think of what could be under that brown water, and the hundreds of crocs I had just seen, or piranha, the penis-eating fish, stingrays, electric eels, or the even worse things I didn't know about. But when I plunged in there was nothing. The river bottom was sandy and without seaweed or leaves, only smooth, uninterrupted sand. The water was cool and refreshing. It was my first lesson in Amazonian Mythbusting 101.

We closed in together, me in front, JJ following. But before we could get within arm's reach of the snake it bolted with impressive speed. "Oh dear!" JJ said in a half-British, half-Spanish accent, and laughed. "I am a little glad he went. I was *poco* scared!" Then JJ bent down and took a big gulp from the river.

"You can drink that?" I asked.

"Oh, *sí*!" JJ said enthusiastically, and cupped his hands to drink more as an example. "Will you try?" Once again the New Yorker in me cringed, but I took a big swig and found it delicious. Lesson two. It was smooth and fresh, and as JJ and I climbed back into the boat he was happy. "Anyone else want to try?" he asked the sitting volunteers, but none took him up on it. "Most people they come and don't drink, but this is the best water in the world," he said, holding his arms out. "Es muy delicioso, no?" I agreed with him and he turned to Emma with raised eyebrows as if to say "not bad." It wouldn't be the last time that a snake encounter would spark something profound.

We arrived at the station later that day and as we unloaded the boat, for the first time in my life I entered a jungle. I fell several times carrying luggage and boxes of food up to the station. I couldn't help myself: it was incredible. When the supplies had been put on the deck of the station Emma instructed everyone to take a half hour to wash up and relax; then we'd have an orientation to go over station basics. I seized the opportunity.

I ran back down the path and to the head of a trail labeled "Link to Transect A." I had no idea what that meant but followed it. Slowly. It was nearly dusk and the twilit interior of the jungle was ominous. I was acutely aware of being alone. It was terrifying and wonderful. Trees loomed so high that it was impossible to see where many of them crowned. There was life everywhere; every inch of everything was extravagantly draped in other things: mosses, lichens, ferns, mushrooms, trails of leaf-cutter ants, and butterflies. There were strange smells and I could hear creatures moving and calling all round me. It was a complete sensory overload. My courage allowed me to go only at a snail's pace down that trail, each step tentative and hushed beneath the ancient trees bearded in moss and vines. I came to a place where the land dropped off steeply to reveal a startling jungle vista. Miles of steaming, savage wilderness stretched before me in the twilight.

No matter the documentaries I had watched, the books I had read, and the daydreaming, I was wholly unprepared for the fantastic world around me. Every atom in my body was humming as I hyperventilated, clutching my forehead. I had never been so awed.

The research station was the most beautiful place I had ever seen. It sat about a half mile from the river, nestled in the middle of the jungle, surrounded by green foliage and decorated with brilliant heliconia flowers. It had a palm-thatched roof, with a kitchen, bathroom, bedrooms, and hammocks to relax in. There was no barrier to the outdoors, and each room was open to the jungle. Geckos, hummingbirds, bats, snakes, and rainbow lizards could all be seen without ever leaving the main deck. In the large rectangular central area was the medical box, hammocks, and library; a cabinet was filled with studies, reports, and field guides, as well as novels left by travelers. There was no electricity or communication to the outside world, and the food we ate came entirely from the sacks of rice and other dry items we had brought upriver, supplemented by produce such as bananas, tomatoes, and papaya from the station's small farm.

Every morning we would wake up at five o'clock to begin research shifts. Sometimes we'd work on mammal senses transects, and other times we would study macaws. Around the stations were several transects, four-kilometer-long paths that had been blazed perpendicular to the river. The goal was to walk as quietly as a ninja, identifying all mammal life encountered, and in recording these observations over time, construct a picture of the local ecosystem. This was where you really got to see some wildlife, as well as Emma and JJ's impressive skill. Back in New Jersey I had been the guy who could find wildlife no one else could, but this was the big league. JJ could predict when spider monkeys were coming, or smell a troop of peccary a half mile out. He spoke of tracks in the mud the way a violinist reads sheet music, interpreting symbols with deft precision, explaining thoughts and motives of each creature while he knelt. "The jaguar came here last night to check for agouti," or "the *huanganas* [peccary] were here for the palm this morning." At first it was all over my head, as was everything about JJ. Once while inspecting the carcass of a peccary killed by a jaguar, he pulled a tusk from the boar's skull and handed it to me, without a word.

At thirty-four years old, JJ was handsome and athletically built, with sharp eyes that were always moving beneath the shadow of black spiked hair. In the forest his intensity was unmatched, and while talking and joking at dinnertime that warm energy was explosive, his smile contagious.

On research transects each morning I was able to observe JJ in what was undeniably his natural habitat. He would often walk miles with no shoes, his calloused feet immune to thorns, nettles, and stinging creatures of the forest floor. While looking toward the canopy he could mimic any animal's call, sometimes carrying on conversations with birds or monkeys, translating to the rest of us. Sometimes he'd cut a piece of bark and hand it to one of the volunteers to sniff, and then explain that it could cure infection or some other ailment. Other times he'd pluck sweet wild fruit from the jungle. Bare-chested, with shoulders slouched and his machete held idly at his side, he'd interact with the forest in a way I hadn't known to be possible. To him the Amazon was not a savage battlefield of dangers and impending doom, as so many explorers had claimed, but rather a bountiful wonderland filled with treasure and simplicity.

Many have documented the curious fact that in the Amazon, despite being surrounded by extraordinary biodiversity, it is possible to walk for days without seeing a creature. The men who recorded such things clearly didn't have JJ as a guide. One day's journal entry recorded the species revealed by JJ's skill in a single morning:

ANIMAL SIGHTINGS:

Spix's Guan, squirrel monkeys, saki monkey, mealy parrots, white throated toucan, saddle back tamarins, giant Amazon squirrel, red and green macaws, peccary (50+), screaming piha, red throated cara-cara, yellow crowned parrot, red howler monkeys (heard them, but no sighting), grey brocket deer, blue morpho butterfly, common swamp snake (two), yellow footed tortoise, leaf cutter ants, and a cane toad.

The macaw research was very different. Instead of walking trails we'd sit inside a mosquito net and count the birds as they screamed at each other and ate mud. Having a macaw colpa by the station was a big deal: it meant that our forest was a mecca for the birds, and that we could easily see them foraging or nesting in the surrounding forest. Most macaw species are vanishing. Though they once ranged from Mexico down through Bolivia, today macaws are in steep decline. This is mostly due to habitat loss and poaching for feathers and food. The Madre de Dios is one of their last strongholds.

Macaws require primary forest to breed. The ones that frequented our colpa were scarlet and red-and-green macaws, both of which roost exclusively in tree hollows, high in the canopy. Most often they use ironwood trees. But finding a large enough tree with the perfect hole is difficult; studies have shown that suitable hollows are so rare that there is usually only one for every sixty-two acres of forest. This limited real estate means that the birds have to rotate who breeds and when, resulting in only 10 to 20 percent of the population actually reproducing each year.

When poachers go after macaws they often don't have the expertise or desire to climb the giant ironwood and so instead they cut it down. This removes one of the rare nest sites from the forest and it could take more than a century for another to fill the gap. In this way the poachers affect not only that family of macaws, but the entire population, and future generations. Thus when forest is removed en masse the effects are multiplied. This is how macaws have become so endangered.

Humans have always been fascinated with the brilliant birds, and a pair of scarlet macaws was among the initial gifts Columbus brought back for the queen as proof of his discoveries—it was the first time the birds were seen on European soil. Even today the birds are coveted as they disappear, and you can find their plumage in earrings and decorations sold in tourist shops throughout Central and South America. The birds are so valuable that the last pair of wild Spix's macaws, a brilliant blue species, was reportedly sold in Switzerland for forty thousand dollars. Now Spix's macaws exist only in captivity, where they are being bred in the hopes of one day reintroducing the species to the wild.

Watching the monogamous scarlet and red-and-green macaws caw and argue on the riverside was mesmerizing. Even after seeing it every day for weeks, the spectacle was hard to comprehend. The green of the forest and brown clay below were sharply contrasted by the tapestry of yellow flowers that fell near the colpa, and once the crimson birds with blue tails arrived each day, it almost seemed unnatural to see so much color.

When research was over for the day, we'd eat lunch and often be free for the afternoon. I would walk the trails by myself, stalking, listening, eager to see everything. I kept notes of every species I saw: birds, mammals, reptiles, amphibians, fish, and even insects. At night I would go out on night walks with JJ, and then prowl around the station with my headlamp even after he'd gone to bed. The next morning at five it would start over again.

I was so wired all day long that JJ was worried.

"You please sleep, no?" JJ asked me in the second week. Everyone else had slouched into hammocks after a long morning's work, but I was only changing my socks before heading back out alone. He studied my face concernedly. "Don't you need to sleep?" I told him I was fine. "You love the jungle so much why? I mean, we have sooooo many people come from everywhere and no one is like this—you *love la selva*, no?" He grinned and stuck a finger under the boar-tooth necklace I was wearing, made from the tooth he had given me from the jaguar kill. I told him about where I had come from, and how long I'd waited to see this place. He squinted as I spoke, and absorbed every word. Then he flashed his wicked smile. "Okay, but tonight you sleep, okay? Tomorrow I'll show you."

"Show me what?"

"You see tomorrow." He said, "Nigh'nigh," and I was left wondering what the morning would bring.

We woke at 3 A.M. for a reverse transect. By the time we had finished at 8 A.M. everyone in the group was beat and trudging back to the station. JJ grabbed me by the arm and told everyone else to continue on. With his machete he pointed into the forest, off the trail, and said, "Let's go."

I had to struggle to keep up with him. He moved like a breeze, fluidly passing through the underbrush, his shirt over one shoulder, machete in the opposite hand. We spent much of the brilliant day walking, and by afternoon the jungle was warm and bright. We traveled without saying a word, moving forward with intensity. I would soon learn that JJ possessed a love of adventure similar to mine, and that he did not joke around about reaching that goal. In those early moments together JJ was teaching me one of his most important lessons about the jungle: if you go where others don't, and spend enough time really *feeling* what is around you, anything can happen. It was an unspoken philosophy that we would take to extremes in the future, but on that sunny day in the forest walking I was green as hell and JJ repeatedly turned to me to say "*Shh.*" Though I was doing my best to keep up, speed and silence are competing qualities to all but the most skilled in the bush.

Every so often, with his finger in his mouth, JJ made a loud wobbling wail that echoed through the jungle. He repeated this in intervals and within ten minutes we heard spider monkeys calling back and tracked them to a fig tree, where they were eating high in the canopy. When we approached the troop was moving slowly east; we followed. For the better part of an hour the long-limbed primates gave us a thrilling acrobatics show. Daring leaps from one tree sent some monkeys through the air for fifty feet before swooshing into the foliage of their target. Mothers with babies leapt nimbly from branch to branch, stopping to observe the humans below, using all four legs and prehensile tail as equal limbs in maneuvering through the canopy.

Moving on, we began following a small forest stream. This stream, JJ said, was one he knew well, and as we walked in silence he stopped often to show me a tarantula burrow, an edible fruiting vine, tapir tracks, or a large catfish. For maybe twenty minutes we hopped and balanced through the small brook, taking in the forest. But upon rounding one bend in the channel, JJ exploded into action. He grabbed my shoulders and pointed straight ahead with his machete. Lying across the width of the stream was a crocodilian more than six feet long.

Seeing any sizable croc for the first time, up close, is like seeing a dragon. Their size and beauty are mesmerizing. Built like armored tanks, the forest caiman in Amazonia is especially formidable. This particular caiman perched above a small waterfall with its mouth open, letting the rushing water carry fish and other morsels into its jagged maw as it basked. In years to come I'd often watch caiman for hours exhibiting this behavior: dozing in a sunbeam for hours, barely awake, occasionally snapping their jaw shut on a fish.

For JJ, who had grown up surrounded by caiman of all sizes and species, this was a mundane encounter. But he wanted to test me. "Catch it! I want to check if he is male or female," he said, no doubt curious how the kid who loved catching snakes would do on a larger reptile. Having grown up watching Steve Irwin religiously, I had daydreamed this scenario many times, usually during class.

We approached the croc slowly, JJ to the flank, and me from the rear. I knew that with reptiles, the neck is the place to grab—but the problem was getting there. "Just grab his tail!" JJ said with a grin on his face and his eyes sparkling more than ever. I had never seen Steve Irwin grab a croc by its tail and knew it was a bad idea, but JJ insisted.

With my heart pounding, I took a quick step forward and snatched the tail, with its sharp spiked scoots. The croc, which had been lying motionless, exploded into action and, anchored by the grasp I had on its tail, came whipping around through the air, 180 degrees. I dropped the tail and fell back just as the foot-long jaws filled with teeth came whipping past my face, snapping shut with a powerful *whack* that echoed through the forest.

The croc sprinted for the safety of the stream, disappearing beneath the surface. JJ clapped his hands and laughed at my shocked face. I should have known better, and in later years would learn to *never* grab a croc by the tail. I had almost lost my face.

JJ leapt into the stream, eyes wide and knees bent, in full action mode. "Come!" he said, beckoning me to join him in the water that hid the reptile. We walked up and down the shallow pool up to our waists, feeling with our feet. Both of us were sure it hadn't made it away from there; we would have seen it since the water was shallow where the pool let out. JJ's eyes were intense and he didn't speak as his feet explored every crevice of the streambed. Then, finally, he looked up with excitement. "Here!" he said. "There is a cave underwater!"

I used a machete to cut a sturdy pole and stepped above the hole to slowly lower the stick inside. At first there was nothing, but when the branch was almost two feet below the surface I felt a shock. A muted *whack* came from below, and the stick trembled and stiffened. Startled, I took my hands off the stick, and it stood there, held by an unknown force.

My eyes met with JJ's as another grin grew on his face. "Pull!" he excitedly whispered. I pulled and lifted the stick upward gradually, until the jaws of the croc emerged above the surface of the water, clenching the stick. JJ launched toward me to grab the furious croc, but it released its grip and fled upstream in a mighty splash. JJ sprinted after it and the croc chose fight over flight and turned on him, jaws agape. The strike missed his legs by less than an inch and he dived to the side—and for an instant the croc's focus was only on JJ's flailing body. I dived then, fully horizontal like a short-stop on a line drive, landing on the croc's back, hands clasping its neck. The moment I landed the croc lashed left and rolled right, throwing me to the side and then launching over me. For a moment I went underwater and then was brought right side up as JJ crashed on top of the croc and me, pinning it.

We shouted in celebration. Lifting the croc from the stream, we inspected the most incredible animal I had ever seen. Its large and threatening eyes were livid, ready to shred either of us if given the opportunity. JJ checked the sex of the croc by inserting a finger into its cloaca. "Male," he said, still grinning. It measured slightly less than six feet. Still shaking in excitement, I asked JJ to snap a photo of me with my first-ever croc, which he did, before releasing the giant reptile.

From then on JJ and I were inseparable. Whenever we could, we'd share research shifts, or go on walks at night through swamps. He was the first person I had ever met who was as enthusiastic as I was for nature, and in the jungle we had our heart's fill of adventure. I was learning so much so fast from JJ, I wrote several times in my journal that my head hurt. JJ began showing me species I had never heard of; we saw things that others in the group never did. Our adventures had a palpable magic, a quality of energy and discovery that was the catalyst of our developing bond. It amazed both of us. Sometimes we'd get back from a hike, and Emma wouldn't even believe the things we told her. JJ said I was good luck with animals and would sometimes shake his head and look ponderously at me, repeating that he'd never seen anyone who loved the forest as much as I did.

JJ challenged me to push myself farther, and it became clear that I really had found home. He showed me how to hitch a ride down the river on floating logs, how to find water in bamboo, what plants were edible, where various species lived, and what to stay away from. We searched for anacondas together, both hoping to find a giant. One lesson involved him deliberately getting me stung by a bullet ant, or *izula*, a two-inch black insect known for having the most painful wallop of any insect on earth. It lived up to its reputation, and I spent a painful twenty-four hours in bed awake, sweating and throbbing with a paralyzed arm.

By the third week I was sleeping out in the jungle every night, suspended in my hammock, listening to the nocturnal symphony and soaking up the smells in the dark. I'd spend entire afternoons climbing strangler fig trees and exploring the canopy. I didn't need anyone or anything when I was in the jungle. I felt whole; alive, turned on, and engaged. It made sense to me. Sometimes I would come back from hours in the jungle, or be heading out when everyone else was settling into hammocks, and could feel JJ watching me.

Once while high in the crown of a tree I found an odd species of ant that had a flat, aerodynamic head and abdomen. I noticed that if the wind blew, the ants would stop running and freeze, so as not to be blown off the tree. But when I'd brush them off into the air, instead of falling to the ground they miraculously would use the momentum of the long fall to navigate back to the tree, lower down on the trunk.

The first two or three ants doing that I took as a fluke. But when I had flicked my thirtieth ant from the tree and watched it deliberately steer back to safety, I knew I had found something: gliding ants. Evolutionarily it made perfect sense: if they fell to the forest floor they'd be eaten, especially in the flooded months of the year when fish are just waiting for insects to fall from above. The ants developed their flange-shaped exo-skeleton so that they could direct aerial descent, and avoid being eaten. I was consumed by the excitement of discovering what had to be a new species and spent hours making detailed drawings and notes on the ants.

Later on I would discover that an insect ecologist named Stephen P. Yanoviak had had the *exact* same experience flicking ants out of the canopy two years earlier, also in Peru. He and other researchers published the findings on the gliding ants, *Cephalotes atratus*, in the scientific journal *Nature*. The fact Yanoviak beat me to it never bothered me. Although I made the same initial observation he did, at the age of eighteen I had none of the tools necessary to describe a species. But the most important tool I lacked was not a Ph.D. in insect taxonomy; it was the confidence to believe that what I had uncovered was something special. I had been conditioned to assume that you are never the first person. Whatever it is, someone has already climbed it, or photographed it, bought it, sold it, biked it, hiked it, and probably posted it on YouTube. It becomes a worldview: don't get too excited because it's all been done before.

When I told Emma about the ants she merely shrugged and told me not to get excited. "Someone or other must have found it at some point," she said dismissively. And even in my own mind, I was so accustomed to the rules of the world I had grown up in that I shrugged it off, thinking she was probably right. In the end she *was* right, but not by much. Only a few years later I discovered something new to science, on my own. That fact, among others, fueled my sweeping realization that throughout my life teachers, principals, society, and so many people around me had told me only what wasn't possible. The jungle was showing me what was.

It was clear, even on this first trip to Las Piedras, that my destiny would come to intertwine with that of the station. One morning near the end of my stay, JJ and I were on Transect A, following a set of jaguar tracks when we heard voices. The LPBS is sufficiently remote that anytime you hear people it means trouble. The moment we heard the voices JJ grabbed my shoulder and threw me to the ground, and from the foliage we observed as a group of men, all of whom were armed, walked up the trail. When they had passed by us JJ silently stood up and stepped onto the trail, making a small whistle that startled the crew of invaders. From the look on their faces it was clear that JJ had seemed to appear from nowhere. My entry was less discreet.

JJ confronted them about why they were on his land and what followed was a tense standoff. I saw for the first time JJ's warm, playful face become twisted and vicious as he spat Spanish back at the shouting men. I couldn't understand what they were saying but the argument nearly came to blows.

That night at the dinner table Emma explained that in recent years their land had been under serious attack. There had been loggers and poachers, and she accurately predicted that soon there would be oil companies as well. The men JJ and I had met were the previous owners of the land, who regretted the sale and simply wanted it back. As she spoke, for the first time I could see the reason for the weariness in her eyes. In her thirties by this time, Emma had spent more than a decade in the Madre de Dios building her dream, and the weight of it was crushing.

It was just Emma, JJ, and eleven-month-old Joseph running 27,000 acres on the Amazon frontier. Though the website Emma had designed brought in a trickle of tourism and research volunteers, it was far from enough to support the taxes on the land, upkeep on the station, and costs of keeping loggers and poachers out. Perhaps if they had a team working in civilization to send them more tourists, volunteers, and business, the undertaking would work, but that was not the case. While in the jungle Emma and JJ had no one to help them attract more people. Listening to Emma speak, I was dumbfounded. If they were having trouble luring scientists, that was one thing, but in terms of tourism I was sure they could make a killing if they got the word out. I told Emma that I would bring volunteers and help them stay afloat. How hard could it be?

I probably said too much too fast, and Emma offered the same token smile parents give to children planning what they want to be when they grow up. She'd heard it all before: everyone who came to Las Piedras and fell in love with the place, everyone said they'd be back. Hardly anyone followed through.

The doubt in Emma's stone-gray, pragmatic eyes was crushing; her expression across the table said it all. After so many years on the frontier she could see through an overenthusiastic eighteen-year-old suburbanite from a mile away; she'd been one herself once. After all, I had spent only a month at Las Piedras and in days would be heading back to school in the United States. She knew I had no concept of what she was up against; no idea about the realities of working in remote Amazonia. The truth was that she and JJ had taken on more than they could manage and were on the verge of real trouble.

4
Jungle Law

Contrary to all justice and reason, in despair they set fire to their villages and fled into the depths of the jungle.
—FRIAR CRISTÓVÃO DE LISBOA, FROM *TREE OF RIVERS*, BY JOHN
HEMMING

Not too far from the Las Piedras Station, close to the Bolivian boarder, another protector of the jungle was also running out of time. As the lieutenant governor of the remote town of Alerta, Julio García Agapito was widely respected. He had earned a reputation as a young, hardworking, brave, and effective archenemy to illegal loggers in the area, a role with a historically low survival rate in Amazonia. But García was no gunslingin' hero; he was a humble Brazil nut farmer and a family man with everything to lose.

In December 2007 he repeatedly filed formal requests for protection with Peruvian authorities and received only eerie silence in response. Like many indigenous people, Julio García saw big-leaf mahogany trees as a source of wealth for his people, one that if properly managed could bring in much-needed revenue to their isolated community. But the loggers had no respect. They came from outside and took everything, with no consideration for rules. They were aggressive and dangerous, massacring wildlife and taking advantage of and even killing local people. Juan had been working closely with INRENA, Peru's resource management agency, to stop the lawlessness and together they had busted numerous mahogany shipments and had some serious close encounters.

On February 28, 2008, a contact tipped off Garcia about a truck carrying mahogany, with license plate WZ-7256. He immediately reported the truck to INRENA, which intercepted it. García's heart must have been pounding as the local police inspected the fresh-cut wood. The truck had been carrying more than seven hundred board feet of timber. The driver had been restrained successfully, but García's mind had to have been racing with the knowledge that there would be retaliation from the loggers. However, he probably couldn't have guessed how rapidly it would arrive.

The INRENA officers were beginning to haul the heavy red wood from the truck when a figure appeared and jumped into the driver's seat. Before anyone could react the truck's engine roared and wheels kicked up dust as the truck sped away with the timber. Juan and the other officers were dumbfounded. Somehow a second driver with a duplicate key had been waiting, as though the loggers had planned it.

The INRENA officers scrambled into their own vehicles and sped off in pursuit of the truck, leaving García at the empty INRENA office wondering how it had all been orchestrated. It was then that yet another figure materialized. Garcia had no time to react. The two men saw each other for only a silent instant before the flash of the attacker's gun filled the room. Ten rounds ripped through him.

The story of García's death spread around the Madre de Dios and the world, even as far as the *New York Times*, but it was nothing new in Amazonia. Three years earlier in Brazil, loggers cornered a seventy-four-year-old conservationist and nun named Dorothy Stang and shot her in the stomach, then the back, then several times in the head. Also in Brazil loggers tied a young girl to a tree and doused her in gasoline before burning her alive, an example to the rest of her community, which had been opposing the loggers. In the lawless west Amazon, standing for what you believe in can come at a price.

In the Madre de Dios and other places in the region, mahogany trees are red gold. The Madre de Dios mahogany boom, as Emma called it, began in the 1980s and reached its climax in the mid-1990s. At that time the demand for valuable tropical timber for luxury furniture in the United States and Europe had already caused a Gold Rush–style siege of loggers to clean out every old mahogany tree from the areas easily accessible from Puerto Maldonado. As the supply diminished, the demand sent the remaining loggers into increasingly remote areas in search of the grand prix of timber, big-leaf mahogany. This led them to the Las Piedras River.

Prior to the mahogany boom, the Las Piedras had been almost completely untouched. Save for a collection of small indigenous settlements in its lower region, the river was virtually uninhabited and pristine. By the mid-1990s, however, boatloads of loggers became a frequent sight; one study reported as many as two thousand logging boats in a single year. The unregulated logging became so rampant across the Madre de Dios that the government eventually outlawed the harvesting of mahogany in all areas except for registered timber concessions. It was an abrupt and drastic measure.

The result was a massive strike in July 2002 that arose in Puerto Maldonado and virtually shut down the city. Government offices were trashed, and stores were shut for several weeks. Amid the chaos at least one police officer caught an arrow in the rear. In November of the same year the illegal loggers elected their leader as president of Madre de Dios Region. Ban or no ban, hordes of loggers continued to pour into the forest. It got to the point where so many loggers were out in the frontier that pimps sent boatloads of prostitutes out into the jungle with tape measures instead of wallets so that they could accept payment for their services in board feet of timber.

The brazen gangs of woodcutters would take what they pleased from indigenous community lands and even entered national parks. It was reported by several park guards that the loggers would fire their guns into the air while passing guard stations in a clear warning to the rangers: *stay inside*. The young and untrained forest guards, paid miserable wages, found themselves in way over their heads.

There were even rumors that the loggers had clashed with uncontacted Indian tribes in the Las Piedras's headwaters. During his time in office, Peru's then-president Alberto Fujimori was denying the existence of uncontacted tribes and claimed that these stories were mere legend. Some people claimed the tribes were real, savage cannibal leftovers of the Stone Age. Others claimed they were ghosts, malevolent spirits guarding the last sanctum of the forest. Only one thing was certain: *something* was up there.

In reality, the people who inhabited the headwaters of the Las Piedras and the Purus Rivers, the Tarahumanu, were the refugees of a previous period of extraction.

From the 1870s through the early 1900s, the rubber boom consumed the Amazonian west. Several decades earlier, Charles Goodyear had made a monumental discovery: the process of vulcanization. The process involved mixing latex with sulfur, which transformed the coarse tree sap into a durable substance, and it changed the face of the industrialized world. The demand for rubber for tires, shoe soles, machinery, and thousands of other items during the industrial revolution in the United States and Europe caused a frenzy. In *Tree of Rivers*, John Hemming wrote, "It made the best gaskets for steam engines. It came to be used in pumps, machine belting, tubing, railway buffers, and later as coating for telegraph wires." At the turn of the century, bicycles were changing the world and the automobile was in its infancy; both needed tires. All of this inevitably resulted in a mass migration to the only place where the invaluable latex could be tapped: the Amazon.

In the fray the Brazilian government designated the Amazon River an international waterway, opening it up to vessels from every industrial nation. For the first time in history the Amazon became a busy highway filled with hundreds of ships from every nation, all seeking to exploit the region's rubber. However, the rubber trees were scattered among billions of other trees within the untouched wilderness of the Amazon. The only way to get at these trees was to walk from tree to tree, from dawn until dusk, collecting latex. Naturally, for the rubber bosses, native labor was the most effective way of collecting the prized sap. The result was a period of genocide and atrocities so savage that the rubber boom ranks among the darkest chapters in human history.

Yet the rubber boom was, in truth, the second wave, an aftershock of the first period of mass destruction in Amazonia.

Whether for spices, metals, slaves, timber, oil, or medicine, jungles have long cast their shadows over the most grizzly acts in human history. It was the white men who brought the Heart of Darkness to Africa, and in the Amazon the Spanish and Portuguese swept across the basin in a slashing, burning, disease-ridden holocaust. Paraphrasing the words of Winston Churchill, John Hemming wrote that "rarely in human history has so much damage been done to so many by so few. A thousand colonists gradually destroyed almost every human being along thousands of kilometers of the main river and its tributaries." One Father Daniel, horrified by what he saw, wrote of the Europeans: "They kill Indians as one kills mosquitoes. . . . And they use—or abuse—the female sex brutally and lasciviously, monstrously and indecently, without fear of God or shame before (their fellow) men. . . ."

Despite the basin-wide slaughter that began in the sixteenth and seventeenth centuries, there remained some tribes that refused to be made into slaves. As foreign enterprises forged into the Amazon during the rubber boom, many tribes moved deeper into the jungle. Already isolated tribes fractured and fled into the most remote and inaccessible reaches of rivers, where the white men couldn't come. When the rubber boom ended, these tribes remained isolated, out of sight and out of mind to the rest of the world. They became the legend of loggers, the only people who sometimes encountered them.

An ex-logger I came to know in later years told me the story of his son-in-law, who had been a logger on the Purus River in the 1990s and encountered a tribe. The loggers had rounded a bend with their motor off, thus surprising an encampment of nomadic hunters from a tribe. The loggers raised their guns, and the tribe raised their bows. No one knows who shot first, but the loggers reported taking down at least one member of the tribe. When their boat had drifted out of range and out of sight of the tribe, the loggers assumed they were safe. Yet on the next bend a silent arrow more than six feet long, with a foot-long bamboo head, tore through one of the men, killing him.

Though reports on Indian fatalities during violent encounters are impossible to confirm, the tribespeople are decidedly the victims of the outside world; it is not they who encroach on us. Their ferocity is what has enabled them to survive, and ferocious they are: one explorer on the banks of the Madre de Dios River was porcupined by thirty-four arrows before having his skull crushed by clubs. For this reason, although they are commonly referred to as "uncontacted," the most accurate term for the tribes that survive in the ranges of Las Piedras and other parts of Amazonia is "voluntarily isolated." They are not leftovers from the Stone Age, but instead a modern people who continue to fiercely defend their way of life.

Thankfully for them and us, the areas the tribes inhabit are far up in the headwaters of the Las Piedras. Down on the lower Piedras, where the station was located, the issue was keeping the loggers from killing all the animals. Logging teams are divided by jobs, with each man having a different position. Some men are cutters, some are cooks, some are transport experts; but *rumbiadores* are the men who strike out alone deep into the jungle to search for large mahogany, for they know the forest best, even among other natives. The rumbiadores also do the hunting.

These expert hunters targeted species like peccary, agouti, spider monkey, howler monkey, tapir, and other mammals for food. But they would often take other species if available.

With large groups of timber bandits traveling up and down its banks, the Las Piedras was put under heavy strain. In the lower reaches of the river you can no longer see large herds of peccary, or spider monkeys, both of which are favorites for dinner in a logging camp. The loggers effectively cleaned out many of the species they hunted, rendering them locally extinct. Along the banks of the Las Piedras and throughout much of the Madre de Dios the greed-driven massacre devastated wildlife. It was this devastation that Emma and JJ had been monitoring and working against at LPBS.

Of the many hunting stories I would hear over the years working in the west Amazon, my favorites were always of those times when the wildlife fought back. One ex-logger told me of a day when he was out hunting on the Las Piedras, in pursuit of what he thought was a tapir. Tapirs are the largest mammals in Amazonia and bulls can grow well over five hundred pounds. Juveniles are born brown with light horizontal stripes, but as they mature they become uniformly gray. Their bodies are hairless and their faces end in a strange prehensile trunk, like a stumpy elephant. Tapir tracks are among the easiest to recognize of any creature in the forest, large and three-pronged, like a sassafras leaf. That is why I have always wondered how the ex-logger could fail to realize he wasn't following a tapir.

He said it was the excitement. The two dogs he was hunting with had caught a scent and sprinted out ahead. For nearly ten minutes a chase endured in which both of the dogs were continuously barking, but as the logger hurried to catch up, one of the dogs let out a pained yelp, then went silent. The chase had led man and animals into the winding trench of a forest stream. When the logger raced to see what had happened to his dog, he rounded a bend and came face-to-face with a full-grown male giant anteater.

As long as seven feet from nose to tail, a giant anteater can stand and look a small man in the face. They have the largest claws in the mammal world and muscular forearms to power them. What the logger came upon was a gruesome scene: one of his dogs had already been gored, its entrails ripped out all over the ground. The giant anteater, he said, snatched up the second dog just as he reached the spot. It lifted the dog off the ground by the nape of its neck and pinched its spine, killing it, and then flung it aside. Next the anteater, still standing on its hind legs, advanced on the logger, slashing at him with five-inch-long black claws like scythes. According to the logger, he barely escaped with his life.

Out of context, the logger's story might sound like exaggeration, but in order to survive in the wild giant anteaters need to be able to fend off jaguars. Any large creature that can fend off the 'pit bull' of the big cats, a muscular 150-pound killing machine, needs to be equipped with incredible speed and the capacity for brutal violence. In 2007 a nineteen-year-old Argentine zookeeper named Melisa Casco was killed when a mother anteater slashed her stomach open in one stroke. In another story from the Madre de Dios, a hunter shot a male anteater, which retaliated by penetrating his abdomen and pulling out alternating fistfuls of intestines. When they found the man's body his rib cage was empty.

Salvador Dali, who was once photographed walking a leashed giant anteater in New York City, wrote that the species "possesses enormous ferocity, has exceptional muscle power, [and] is a terrifying animal." In the regional parlance of the Madre de Dios giant anteaters are called *oso-bandera-gigante*, which literally translates to "bear, flag, giant." The name is evidence of the confusion that ensues in trying to describe an animal with the strength of a bear, a pluming flag tail, and tremendous size. The giants are alternatively called *tamandua gigante*, borrowing the name of the smaller arboreal anteater and just adding the adjective *gigante*.

Giant anteaters are the largest of the "Amazon claw" family, which includes sloths, armadillos, and anteaters. They walk on their knuckles to spare their valuable claws, which are used to fend off jaguars and to excavate ant and termite mounds. Their claws are a heavy-duty defense to compensate for their slender, fragile head. They have poor eyesight, a keen sense of smell, and no teeth. Along with their size, local legends surround them because the hind footprint of a young anteater is almost identical to that of a human child, with arch, toes, and heel, leaving delicate impressions in the ground. Because of their mystique and the fear that comes with it, fearful farmers who worry about losing their dogs, or their own lives, often kill them. It was exactly this scenario that unfolded on the Piedras in 2007.

Just months from the time that Julio García Agapito was gunned down near Alerta, and in the same year Melisa Casco was gored by the anteater, I watched as a team of loggers drifted down Las Piedras. Our boat was also traveling downriver but much faster than the loggers', and as we passed I could see that their hull was filled with the carcasses of many animals. I was glaring at them, lost in thought as we passed by.

"Ay! Watch out!" JJ shouted, motioning to turn sharp right. He had also been watching the loggers and hadn't realized that we had arrived. I leaned over the edge of the boat, pulling the motor's steering bar toward my chest as we raced toward the shore. With a flat palm pumping the air JJ motioned for me to slow as we neared the bank. Then his palm sliced horizontal: cut the engine. I did and JJ hopped barefoot onto land as the boat gushed into the soft clay. "Bien hecho," he said, smiling; *well done*. I was coming along decently as a *motorista*.

Following my initial expedition on Las Piedras I had returned to the United States transformed. The anger and frustration I had built up in high school had dissipated in the jungle. I had returned home with photos, feathers, and stories I was eager to share with anyone who would listen. My relationship with my parents was back on track, and commuting to college allowed me to continue working and saving so I could concentrate on the thing that mattered more than anything else: getting back to the jungle. Though I had spent only a short time with Emma and JJ at Las Piedras on that first trip, the way I returned home you would have thought I was a Ph.D. who had studied in the Amazon for a century. I read the great explorer-naturalists like Wallace, Bates, Sprice, Shultes, and Munn; I studied species guidebooks. And I had difficulty carrying on a conversation about anything that didn't relate to the Amazon.

In my time at home I finished a semester at school with passing grades and managed to keep my word to Emma and JJ. Determined to join the fight to save Las Piedras, I had used my time at home to recruit enough volunteers for an entire expedition. It didn't matter that they were all my friends: it was a group, and that is what JJ and Emma desperately needed. I was proud to be helping. Emma and JJ were the kind of people I had always dreamed of meeting as a kid, heroes bravely protecting limitless wildlife in the jungle—people who stood for something.

JJ and I picked up right where we had left off months before, as student and teacher, and as friends. Together we brought the group on the two-day journey to the station, and settled into macaw observation and mammal transects. But this time the expedition was very different: Emma was in the United Kingdom with Joseph, which left just JJ and me to run the trip. I was now a guide. JJ was overjoyed that I had followed through on my promise and that I had retained everything he had taught me the first time around. He treated me like a second in command, a position I was proud to fill.

JJ held the bow rope as the volunteers, including Noel and other friends from back home, disembarked onto the beach. Just days into a two-week expedition we had brought the group downriver from the station to investigate a local farmer who, it was rumored, had shot a mother giant anteater. The farm was small, several acres that had been slashed and burned the previous year. In the cleared area were three small huts, a pigpen, and a field where corn and yucca grew. The jungle stood patiently at the limits of the field, waiting to reclaim what had been taken from it.

In keeping with Peruvian etiquette, we were not able to get right down to business and ask the guy about the anteater, but instead had to act as if we just happened to stop by. JJ told the farmer and his wife that we had come to explore, and after they fed us a meal of fried peccary and rice, the farmer took us to investigate a nearby lake. He showed us the carcass of a mother black caiman that he had killed after it defended its eggs when he and his dogs had gotten too close. The croc had been more than thirteen feet long. The farmer also showed us the skeleton of a twenty-foot anaconda he had shot; it was also a female. "This asshole likes killing females, eh?" Noel asked me as we hiked. After spending our childhood adventuring in the woods back home, naturally Noel was among the first to come with me to Peru.

That night we slept at the farm. JJ spent hours chatting with the farmer, exchanging stories, and tolerating the farmer's pain-in-the-ass kid, who wouldn't leave him alone. First JJ asked the kid to stop bothering him, but when the boy wouldn't let up, JJ got angry. What can you do against a seven-year-old? When the kid ran by and yanked his hair, JJ had had enough. "You wanna play?" he asked with a twinkle in his eye. The idiot kid thought it was a game and said yes. JJ lifted the boy above his head and told him to hold on to a beam in the roof, which the child did. Then JJ walked away. Stranded and terrified, the obnoxious child burst into tears, holding on for dear life until someone heard him and came running. After that the kid left JJ alone.

The next morning we awoke to a filthy yard bustling with chickens, dogs, cats, pigs, and ducks, that was covered in feces from all. While JJ and I prepared breakfast for the group, it was impossible not to notice one of the farm's dogs. It was lying still, bleeding.

We found out that the previous day, while hunting with its owner, the dog had chased a herd of peccary. The boars had surrounded the disoriented dog and plunged their fangs into her body. As we inspected her, there were dozens of gaping three-inch-deep holes all over her body, brimming with squirming maggots. I felt the color drain from my face, and I could see Noel's hands trembling. We had never seen such hopeless suffering. The pain the small dog was in was heartbreaking; it could barely lift its head. The farmer laughed and said he expected her to die within the week. I asked the farmer why he didn't just kill the dog and put it out of its misery. He said he didn't want to waste a bullet.

Noel and I talked and agreed that there was no way of saving the dog, and that to leave her alone to die slowly over the course of days was inhumane. We asked the farmer if we could put her down. The mother murderer didn't care either way. So with heavy hearts Noel and I carried her into a field, under the shadow of the jungle beyond. I watched as he petted her, fed her, and made her last moments as comforting as he could. Then he stood and nodded to me. I held her head, whispering and stroking reassuringly as Noel brought an ax high over his head. The stroke cleanly passed through flesh and bone and into the earth. Death was instantaneous. We buried her and returned to the farm, where at last JJ got around to asking about the anteater.

The farmer's wife was excited to show us her newfound pet, and called out into the corn field: "Luluuuuuu." We all looked at each other incredulously. There was no way an *anteater* was going to come when its name was called; but stalks of corn rustled and within moments a giant anteater emerged. "Mira mi tamandua pequeña!" she said. Look at my little tamandua.

She was tiny, roughly the size of a beagle, with a tail as long as her body. Her bristly coat was gray and she had a thick Mohawk of white along the ridge of her back. Decorated with black bracelets and triangular flank markings, she was a beautiful animal. Trotting from the corn up toward our group, she ran straight for me. Simply out of some habit from beckoning my own dogs, I bent down to her and held out my hands. She stood on her hind legs and spread her arms. Again I reacted as if on autopilot and lifted her from the armpits like a child. She turned her long alien face toward mine and slurped a ten-inch tongue across my cheek. Everyone laughed, and someone took a picture. Noel and I looked at each other like *What the hell is this thing?* Even JJ had never seen a giant anteater so close-up.

As our group fawned over the bizarre animal, the farmer's wife explained that for the last month or so she had been feeding the anteater powdered milk mixed with water. The woman had used an old plastic container and stretched the spark-plug protector from a chain saw across the opening to form a nipple. She demonstrated how she fed the baby giant, and we all watched spellbound as Lulu curled a black claw over the woman's hand, holding on as she ravenously gulped the milk.

In the end it took almost two hours of slow, tactical conversation on JJ's part to convince the woman to let us take the anteater. We were sure that if it stayed, her husband would sooner or later dispatch it; that is, if the dogs didn't get it first. In the end she agreed, but only after a flashlight and half the contents of our boat's gas tank had been parlayed. The fate of the tiny orphan was now in our hands.

5
The Giant

God is really only another artist. He invented the giraffe, the elephant and the cat. He has no real style. He just got on trying other things.

—PABLO PICASSO

She was a monster. On the boat trip upriver Lulu had quietly latched on to Noel's thigh and barely moved for two hours while he broodingly turned the pages of *Heart of Darkness*. The rest of us had made the hasty judgment that this animal was calm and sweet. Now back on the main deck of the station, though, she had become a terrorist.

Upon returning from two days of camping, everyone hit the hammocks, which development Lulu seemed to find infuriating. As soon as she wasn't in someone's arms she ran toward the hammocks, her claws clip-clopping on the wooden deck. Galumphing up to my buddy Ben's hammock, she stood on her hind legs and sank her claws into his stomach. Ben screamed and rolled out of the hammock as her razor claws sliced fabric and skin. Still on her hind legs, the anteater wheeled around and then chose her next victim. She made for JJ and slashed at him as well. JJ jumped to his feet with wide eyes and asked, "Que pasa, tamandua?"

If you bred a hyper baby black bear with Edward Scissorhands, the result would be something similar to what we were dealing with. Though she was small, her claws were already three-inch-long black sickles that could tear through denim and skin with ease. She went from hammock to hammock attacking and slashing. Everyone was jumping, running, and laughing amid the onslaught. She seemed to be after anyone she could skewer. Here was the weirdest animal any of us had seen, something strange enough to be a creature from *Star Wars*, advancing angrily toward whomever she could find and slashing them to ribbons.

When she came toward me I curled into a ball on the floor in mock terror, knees down, hands over my head. Lulu approached and stood on her hind legs, and a worried Ben warned, "Dude, be careful, she can really cut you, man!" I didn't have time to move before she was on me. I felt a claw latch onto my shoulder, and then another on to my spine. For some reason, despite the pain, I stayed still and Lulu pulled herself onto my back, where she grunted, wiggled, snuggled in, and seemed to fall asleep instantly.

Lifting my face from my hands, I looked up to see everyone staring, stunned. Suddenly it made sense.

Female giant anteaters give birth one at a time, and share an intimate relationship with their newborns. The infants spend the first nine months of life on the mother's back. Lulu almost certainly remembered holding on to her mother's fur, riding through the jungle in safety. Though they begin eating ants after eight to twelve weeks, they continue riding and learning from mom for months. Only weeks ago Lulu and her mother would have spent nights in the jungle curled tightly, newborn nestled in the center, both blanketed by the mother's thick tail. For the last few weeks, however, she'd been on her own.

We would learn that she *needed* to be draped over someone's knee or lying on somebody's chest at *all times*. If we neglected this need, she'd claw and scream until she got what she wanted. As a result, in the days and weeks that followed, I was rapidly forced to play the role of full-time mother anteater.

On that first day, though, I carefully coaxed the tiny anteater from my back to my chest and sank into a hammock, shrugging to JJ that it seemed like I had no choice. He grinned understandingly and said he'd take care of dinner and other things for the night. Lulu grunted and purred as I stroked her coarse mane, eyes closed tight, seemingly soaking in the thrum of my rib cage on hers. Together the anteater and I slept. It was the start of a rare and profound relationship between human and animal. We spent the night fast asleep together in the hammock and the next morning went like this:

I woke behind closed eyes and wondered where I was for a moment; something was shaking me. Cracking an eye open, I saw the elongated snout and eager eyes of Lulu inches from my own face. She had slept the entire night on my chest but now was ready to play. She could see that I was awake and wiggled with happiness, shot several quick slurps across my face with her tongue, and fixed her large front claws into my ribs to gain better position.

I howled and tried to remove her black claws from my sides as quickly as possible. As my hands were busy with her claws, she capitalized on my vulnerability and dug her little black snout against my exposed ear and sent rapid-fire laps from her ten-inch, ant-grabbing tongue into my cranium. What a way to wake up.

I lifted her off my chest as a half dozen more whacks of the tongue hit. With hind feet on my stomach and front claws safely around my thumbs, I held her in a semi-standing position, careful that my face was just out of tongue's reach. She bobbed her head and flung her tongue in all directions, seemingly thrilled with herself after a well-staged wake-up attack. I placed her on the deck and she stretched before trotting into the kitchen, where she could hear the others preparing breakfast and smell her bottle being mixed.

Over the next two weeks JJ and I continued to lead the expedition, and everyone at the station learned and fell in love with the little anteater. We all took turns feeding her, napping with her, and playing with her. But when the expedition ended, it was time to make some decisions.

Originally the plan had been for JJ to take everyone downriver and I would stay at the station to care for Lulu. In another two weeks JJ and Emma would be back with more volunteers. That meant I'd be living in the jungle alone with the anteater in the interim. But feeding Lulu each day had depleted our entire milk supply, and I'd coincidentally caught a nasty fever. In the end I opted to travel downriver with everyone else to Puerto Maldonado to get medication for me and milk for Lulu, and while I was there, to see my friends off. But I was left with a serious problem: how to get back to the station.

On a boat with JJ, Emma, and a group of volunteers, Las Piedras was an exhilarating river to navigate. To travel it alone, at just nineteen years old, when it was all so new and still foreign, was another story.

We left Lulu with a native guy we'd hired named Pedro, and made for Puerto. But in town JJ was busy running around doing legal stuff, trying to fend off the armed men he and I had run into months ago. He had taken the matter to court and was at a crucial stage, which meant I was on my own. JJ and Emma's struggle to maintain Las Piedras had put them in the red in terms of cash, and their decade-long relationship had become similarly tense, fraying under the strain.

My mission, however, was focused on one thing: caring for my little tamandua. The day after we had come downriver, after all of my friends had flown out, my fever was mostly gone and JJ dropped me off on the banks of the Madre de Dios River, across from the Las Piedras's mouth, and wished me good luck. It was my first time traveling alone through the jungle.

I spent hours feeling obtuse as passing gold miners and loggers stared at me. I tried to find boats with families, and would use the full extent of my Spanish to ask them if they were going "arriba Piedras," up the river. It took several hours of trying but eventually I found one group who waved me aboard. I hadn't fully beaten the fever and was feeling sick and a little scared as the boat left the world behind and was swallowed by the ominous jungle mouth of Las Piedras. Something in me didn't feel right. I admit that I repeatedly thought of my anteater, and told myself insistently that everything would be fine. I was too young at the time to know that my gut is almost never wrong when it comes to sensing doom.

At first the people on board eyed me cautiously, not knowing what to make of the stranger in their midst. The women kept their children close and the men were guardedly stern. They must have been wondering why this gringo would be going into the jungle where there were no hotels, stores, roads, or anything really. Was he *loco*? It seemed like the entire crew did nothing but stare at me for the first hour of travel, until I bent over the side of the boat and filled my water bottle to drink. Instantly they began whispering to one another. Repeating what JJ had always said after drinking river water, I turned to them, smiled, and said, "Que rico Piedras," essentially: *delicious Las Piedras water!*

Once they saw me drink, they all smiled and the tension broke. In years to come, intentionally drinking from the river, and making a show of loving it, was a first step many, many times in gaining the trust of local people. Because everyone in the Madre de Dios knows that as a rule the gringos don't drink the river water, when one does, the local consensus seems to be that he or she must be all right. The truth is that back home in New York I often long for a gulp of Amazon water. On the boat that day I kept thinking, *Thank you, JJ!*

The atmosphere on board the large canoe changed entirely. Inquisitive children inspected my bag and person, and one woman with kind, hooded eyes offered me salted bush meat for lunch. The family was made up of two husbands, two wives, an old woman, and half a dozen kids; the guy driving seemed to be an uncle. Though they didn't speak a word of English and I knew only a few words of Spanish, we communicated well. The kind-eyed woman and her husband seemed entertained by me. They were as curious about me as I was about them, and they asked me many questions, including why I was going up the river. It must have seemed strange to them. I could only smile. They shared their food with me, mostly rice and yucca, and took great pleasure in pointing out wildlife. At night we camped on a beach beneath the stars.

We traveled all day the second day, and the second night it was already dark when we pulled ashore. The uncle had been driving and nodded in approval when I hopped off the front of the boat to tie us to shore, again demonstrating what JJ had taught me. I walked a few paces along the beach to pee as the family clamored in the boat; I assumed they'd be there when I got back.

When I walked back to the boat it was gone, and so were the people. So was my backpack. For a long moment I stood in the dark, dumfounded. It was one of those moments when you doubt what your eyes see. Had I missed something, or did I just get robbed? The river rushed by and the jungle towered above, indifferent to my turmoil. I was nineteen years old and alone in the Amazon. I should have known something like this would happen. "Hello?" I called out, but no one replied. For about ten minutes I stood in complete shock, in complete darkness.

"Señor!" a voice called out. I turned and could make out the shape of a child; it was a girl no more than ten years old. She was barefoot and naked save for something around her waist, and held out her hand to me. Still dumbfounded, I took it and followed. She led me through the jungle on a path, though I don't know how she saw. It was pitch black out. She held my hand the entire time as I stumbled along, fireflies and biolumi-nescent fungus glowing in the darkness like alien Christmas lights. We walked for almost ten minutes, far from the river, before I was able to make out firelight ahead.

We emerged into a clearing where a dozen thatched huts stood close to one another. In the center was a fire, around which several people sat cross-legged on the ground. As we walked through the huts, sparsely illuminated by the firelight, the village enveloped us. Beings moved in the shadows all around and from hollow doorways I felt the curious stare of unseen eyes. The little girl led me to the back of one hut, where

I recognized the family I had traveled with. One of the men walked up to me, gave a hushed chuckle, and patted me on the back as if to say *there you are!*

It was barely nine o'clock but to people who live by the rising and setting of the sun this was the middle of the night, and he pointed to where I could sling my hammock. They had also carried my backpack from the boat. Once I had climbed inside, the kind-eyed woman from the boat who had offered me lunch approached. She advanced with an expression of sage understanding and placed her palm on my forehead for a long moment. Smiling warmly, she said something in Spanish before turning to go to bed. The motherly gesture was comforting; it was as if she could see how out of place and thinly stretched I was.

For a few hours I stayed awake, too awed by my surroundings to sleep. I watched the group around the fire, illuminated in orange light. This was a village that looked as villages in Amazonia must have for thousands of years: split-bamboo walls, palm thatching, and balsa cord ropes and hammocks; water buckets made from gourds were placed by roof corners to collect clear rainwater. Monkeys lay curled in the shelter beneath the thatched roofs, and more than one blue-and-yellow macaw slept quietly on some perch or another. Built on stilts to escape the floods of the rainy season, small huts for individual families were scattered about, with the occasional communal longhouse. Everything was entirely from the jungle.

The chorus of frogs and insects in the jungle night was omnipresent and overpowering, playing in pulsating waves as an old man sang softly. Huddled in my hammock, I took in the dark world, acutely aware of my surroundings. Traveling alone through the Amazon on a mission to save a baby giant anteater . . . If this was not the adventure I had always dreamed of, nothing was.

The following day, before dawn, we left the village and battled the rapid current of the river until evening, when we arrived at La Piedras Station. I hopped off the boat and thanked everyone aboard. They had taken me on a days-long trip, shared their food, and cared for me. As the boat pulled away, leaving me onshore, they waved and smiled warmly, and I did the same, dearly wishing I knew the words to thank them. Once they had gone, I ran along deserted trails toward the station through the jungle night and found Lulu in her box. After days apart we embraced and spent the remainder of the night in a hammock dozing.

The next morning Lulu woke me up with her signature attack, and I cringed as my morning brain-cleaning came in rapid slurps. Then she and I set out exploring. Being at the station without Emma, JJ, or anyone else was incredible. I was feeling as good as ever after surviving my journey upriver and being reunited with my anteater, and was glowing with the reality that I had two weeks ahead of me to simply soak in the jungle and enjoy my strange new friend.

On that particular day we had walked for hours on and off trails, at one point encountering a large group of spider monkeys moving through the treetops. I removed my shoes for increased stealth as Lulu and I stalked below the troop of long-limbed primates. They were calmly foraging in the crown of a large strangler fig when the rumble of peccary registered to my ears. Crouching, I held the anteater close. As Noel and I had learned a few weeks earlier from the poor gored dog, peccaries can be dangerous.

What we were about to witness was an interaction between several key species of the Amazonian landscape: spider monkeys, peccaries, strangler figs, and huicungo palms. The strangler fig that the spider monkeys were perched in is one of the strangest trees on earth. A gigantic and deadly species, figs occur in many tropical regions, and in many locations they are crucial for the surrounding ecosystem. The reason they are so integral is their value as food, shelter, and structure within the forest. They produce fruit at odd times of the year, like the dry season, when most other plants are barren. Out of more than fifteen hundred species of plants, only about a dozen, or roughly 1 percent, produce fruit during the driest months, and none in greater abundance than figs. These select plants support everything from minuscule birds to twenty-pound spider monkeys, fruit bats, and myriad other arboreal species. These and other animals can get as much as half of their diet from figs, depending on the time of year. But it is the fig's reproduction that is its most fascinating trait.

They are pollinated epiphytically, which means that they use the branches of other trees like flowerpots to grow from. The jungle is adorned with thousands of varieties of epiphytic plants, orchids being one, that grow in this way. When a monkey or bird defecates after eating the fig's fruit, they often deposit seeds in the crown of a tree. Epiphytic plants latch on and use the collected rainwater and decomposing leaves to survive. But the strangler fig is not content simply to live on another tree; instead it wants to overwhelm it.

From its perch the seed begins to send down roots. They grow from the top down and thereby shortcut the struggle for light that almost every other tree species endures in the jungle. These roots travel down the host tree's trunk until they reach the forest floor. The organism then kicks into high gear and begins attacking its host tree from above and below. The long, slender roots grow in girth and multiply in number, surrounding the trunk of their host, while above, the fig begins to slither out onto the branches of the existing tree to unfurl leaves of its own that outcompete those of the original tree. The roots of the fig by this time are so massive and numerous that when the host tree rots and disintegrates, they maintain structure. Mature figs stand as hollow columns stretching up to the canopy. Tangled and vast, they provide habitat for innumerable species of mosses, lichens, insects, reptiles, amphibians, birds, and on and on. On this day it was spider monkeys.

The monkeys in the strangler fig were clearly intrigued by the approaching army of boar. Emma had explained that peccaries are extinct throughout many of the regions where they once existed, including Las Piedras, because they are so prized by hunters and hungry mahogany loggers. So having herds of one-hundred-plus individuals inhabiting Las Piedras was an achievement. Lulu seemed to have the good sense to stay quiet as the pigs approached. The rumbling of the earth as they passed was something akin to the vibrations of a freight train. The bass of their grunts was resonating in my chest, their tooth clacking and the squeals of the young echoing through the forest.

Thankfully the hogs moved past us, to below where the spider monkeys were feeding. The peccaries were interested in feasting on the strangler fig fruit that the monkeys were dropping. They were also chomping on the fruit of huicungo palms. Huicungos are very different than strangler figs: they are a small- to mid-sized palm in the Astrocaryum family that dominates the interior of the canopy so profoundly that it is the second-most abundant tree in much of its range. The key to its success lies in its ability to grow well in low light, and the armor of savage spines that protects them. From two to seventeen inches they jut from the tree's main trunk in the thousands, needle sharp. They are jet black and perfectly straight, and so thin that even the slightest encounter will send them plunging into your body. This defense makes them impossible to approach, let alone climb, and keeps them safe during their slow growth. Over the years I have seen dozens of people get skewered through their shoes, and on a few occasions seen a spine completely impale a finger or foot. It is not uncommon for me to work tips out of my feet even back home in the United States, months after leaving the field.

As the monkeys moved through the canopy, dropping figs and huicungo fruit, the peccaries feasted. It is in this way that the large black primates and bombastic pigs help shape the landscape of the jungle; without the monkeys and peccaries, many trees would not have the opportunity to reproduce successfully; without the trees, the monkeys would have no food and no home. But it's not solely about growth; it's also about control: studies have shown that in areas where peccaries have been hunted out, the nasty huicungo palm spreads uncontrollably, dominating and altering the composition of the forest.

John Muir famously wrote, "When we try to pick out anything by itself, we find it hitched to everything else in the universe," a statement that in the Amazonian context is more immediate reality than philosophical musing. As biologist John Terborgh put it: "Subtract figs from the ecosystem, and the whole thing can collapse."

After Lulu and I had watched the monkeys and pigs for some time, the monkeys seemed to become enraged at the hogs below and started showering down sticks and fig fruit. The peccaries roared with insult and crashed about, making as much noise as possible as they looked up at the monkeys. Some of the lanky-limbed primates came surprisingly low in the trees for better accuracy, and launched whatever ammunition they could find at the black pigs, while cursing loudly. In years since I have never seen the two species interact in this way.

My observation of the scene ended when a large branch snapped under one monkey's weight and spun onto the ground with a powerful *thwack*. I had placed Lulu on the ground by that point. She reacted in terror to the nearby impact and bolted for the safety of my arms. I was still looking up with wonder when she leapt off the ground and onto my leg, her black claws piercing the denim and flesh. I yelped in pain.

The peccaries realized how close I was to them and ran. Stampeding in a din of crashes and grunts, they barreled off into the jungle, while above the monkeys retreated to higher branches. Our cover was blown.

Part of my job every day while living with Lulu was not only to make sure she was fed, but also to encourage her to eat ants. I had no idea if this was something she would do naturally, or if it was something her mother would teach her.

In the Amazonian ecosystem, ants alone can make up as much as a third of the total biomass of the landscape, roughly 30 percent of all living organic matter. That total includes everything in the ecosystem from termites, bees, and flies to mammals, and even trees; everything alive. It is no wonder then that giant anteaters are so at home. In a strange way they are similar to humpback whales: both species have unique physical adaptations that allow them to exploit the surplus of tiny invertebrate caloric opportunities that are off-limits to other carnivores. While a jaguar has to fight and kill for almost every calorie earned, the anteater has the luxury of jogging through the jungle, sampling the endless buffet. A giant anteater's diet requires a raw protein intake that resembles the dietary needs of an obligate predator (i.e., animals restricted to a carnivorous diet), like a jaguar, tiger, or crocodile. But they do not have to "hunt" and don't even have teeth. Instead they have one very special tool, an evolutionary key to the city that grants them access to rapid arachnid consumption: their tongue.

A giant anteater's tongue can be as long as two feet, extending deep into the ground in rapid slurps, as I learned from Lulu's brain-cleaning morning wake-ups. They have specialized firing muscles and active salivary glands that power the tongue and keep it sticky. The combination of their excavating claws, long nose, and even longer tongue makes it difficult for ants and termites to hide. They are voracious eaters—which is why it scared me that Lulu only seemed interested in milk.

The plan was to rehabilitate her so that she could one day live in the forest, but even in my care her future was far from certain. For one thing, back in those days I knew nothing about what baby anteaters need to survive.

With no communication to the outside world, I could only hope that anteaters produced milk that was roughly the same percentage fat as the crappy little bag of evaporated milk that we had bought from Puerto Maldonado. The chemistry of milk between mammals can vary greatly. Species like hooded seals produce a 65 percent fat content, while some whales produce 45 percent. To put that in perspective, humans have a 4.5 percent fat content and cow's milk has between 3.5 and 5 percent (depending on the breed). The purpose of milk in large mammals is to give young the high-fat and high-energy boost to go from infancy to efficiency as rapidly as possible. If the milk I was giving Lulu didn't meet her needs, I worried, she might not grow properly, or worse, might not survive at all. Anyone who has cared for orphaned wildlife knows that the odds are always against success.

I knew that I would have to work carefully on Lulu's diet if she was going to survive, let alone grow into the six-foot giant she could one day be. I wanted to get her on ants as soon as possible. To make it easier for her to get the protein she needed, I made mixtures of milk, mashed bananas, and crushed ants. These she always accepted with joy. It looked like a revolting ice-cream shake with squirming black sprinkles.

I wanted her to be able to eat ants on her own, though, and recognize them as a food source, and so I would take her on walks through the forest, not as a human, but as an anteater. On my hands and knees alongside Lulu, I plodded through the jungle. She was noticeably happy to see me at her level, and stayed close. From my ground-level vantage point I was afforded a chance to experience the way the little tamandua took in the world. We traveled locally, one of us looking utterly ridiculous, for some time. At the base of one tree was a large brown termite nest that appeared to be the perfect place for a first lesson. Crawling alongside it, I made a show of sniffing loudly at the mass and inspecting it as an anteater would. Lulu joined in with enthusiasm, poking her long snout in under mine.

Inside the smooth exterior of the nest were tens of thousands of termites, calmly patrolling the labyrinthine passageways of the mound. They had no idea what was about to hit them.

I took two fingers and bored into the wall of the mound, instantly exposing several hundred scurrying bodies. Lulu didn't have to be shown what to do next. Latching her long black claws deep into the mound, she tore it in two. It was as devastating as Godzilla destroying Tokyo. As the termites scrambled in pandemonium, I pictured the event from their perspective, looking up as the nose of a monster swept overhead like a vacuum of death; slow-motion giant sounds as an alien tongue smashed and slapped over the tunnels and byways of the once-peaceful termite city, snatching larvae and soldier alike and whisking them upward.

With surprising speed her tongue was shooting in and out in slurps as she swept the exposed masses, and also snatched hundreds from deep within tunnels. The termites didn't have a chance. This, however, only lasted briefly. Within less than a minute she seemed to lose interest and scurried off. Throughout the rest of that day and the studies of days to come, it was a matter of concern that she never seemed to gorge on ants. But she slowly seemed to be gaining an appreciation for the protein.

I figured that my best shot was reinforcement, and so spent entire days with that little anteater, visiting termite and ant mounds. Sometimes I would spend hours on my hands and knees, with Lulu on my back, just moving through the jungle. I got the feeling that we confused the other creatures of the forest. One day a toucan swooped low onto a branch and cocked its head in our direction, skeptically studying us. Another time a crew of squirrel monkeys came above us to chatter, and I swear it sounded like they were laughing.

Together we spent a week of uneventful yet vividly remarkable days. Lulu and I continued our long walks, and I continued to learn about the forest. During this period we saw Pedro from time to time. Often in the morning he was off in the bush, but in the afternoons we would meet. More than once we loaded Lulu into the small dugout canoe and spent the afternoon fishing on the opposite side of the river. Other days the anteater and I would go solo, staying deep in the jungle to our heart's content.

It was life as all life should be. Living close to the land, we had beauty and adventure all around us. Liberated from the soul-crushing repetition of scheduled existence, every day was new. Far from the distractions of technology and responsibility, each moment was ours. We were never idle. Collecting fruit and vegetables from the garden, mending the thatched roofs of the station, or fixing fallen beam bridges on trails, fishing, cooking, even bathing—all became rewarding experiences. Life was simple, and accomplished with the most basic of tools: bare hands and machetes. These were brilliant days.

In hindsight, my time at the station with Lulu and the years that surrounded it were an age of innocence. My horizon contained nothing beyond the continuation of my friendship with JJ and work at the station, and above all else the quenching of my seemingly endless desire to listen to the jungle's teachings. There was nothing more important in my world. In my mind's filing system it is a chapter tucked in a folder of sunlit green warmth. Years later, beneath slate skies and blackened canopy, during times of grave struggle, when the stakes had risen too high to bear, I would long to return to that time of carefree, elemental simplicity.

The jungle was gradually coming into focus, and what had once been an unintelligible mess of green began forming into recognizable elements. My extraordinary luck in the jungle seemed to have only intensified with time, and animal sightings highlighted our days. I would spend hours watching troops of spider monkeys, or cautiously following in the footsteps of jaguars. These and many other species were no longer exotic wildlife, but a type of familiar neighbor. All of the creatures became this way. Long afternoon naps cradling Lulu contrasted with rugged expeditions into uncharted bush; our activities almost always culminated in a long session of splashing and frolicking on the shores of the river at dusk. In short, Las Piedras became my Walden. I was beginning to build on what JJ had taught me and find the innate sense of the forest that only comes through quiet observation.

Sitting with Lulu cradled in my arms, I'd admire her beauty and listen to the twilight chorus of the jungle. The black bracelets on her front limbs and a large triangular mantle of black were framed in a white border on her flanks, and that big bushy tail. Alone in the jungle at the candlelit station, she'd lock her front claws over my wrist and her tiny mouth over the makeshift nipple to drink her evening bottle of milk. Several inches of tongue hung out off to the side as she closed her eyes in happiness, all the while grunting and slurping. I remember studying her as I lay in my hammock, soaking in every visual and tactile detail of that incredible alien wild animal I had come to love so much. I had no way to know that very soon we'd be pulled apart.

6
The Basin of Life

How many hearts with warm, red blood in them are beating
under cover of the woods, and how many teeth and eyes are
shining? A multitude of animal people, intimately related to us,
but of whose lives we know almost nothing, are as busy about
their own affairs as we are about ours.
—JOHN MUIR, YOSEMITE VALLEY, 1898

The jungle at night is the greatest freak show on earth. When the sun goes down the landscape welcomes a churning nightshift of murdering, slithering, creeping, fornicating, stalking, swimming, glowing life. To walk the Amazon by night is to enter a world where you are gravely disadvantaged compared to millions of sensory savants. Some exchange chemicals signals, others complex vocalizations outside our auditory spectrum, and almost all have night vision. JJ and I had spent hours out at night with headlamps, searching for anacondas and other creatures, but alone it is something else altogether. One night I put Lulu to sleep at the station and headed out on my own into a world that few experience.

In the rainy season the sound of frogs and insects was

deafening, so loud at times that I could feel my eardrums vibrating uncomfortably. Millions of mist particles swirled in my headlamp's beam as I stalked slowly, timidly, down the Brazil nut trail, about two miles from the station. I tried to make each step silent, because with limited sight, sound becomes the most important sense. Somewhere in the distance a crested owl gave a menacing growl. Nearby, a bamboo rat barked repeatedly. There were other sounds, too: rustling and cackling volleys so deranged I couldn't guess what it could be. Mammal? Bird? Insect? No clue. My pulse was spiked and I was on high alert. As if it weren't enough to be living isolated from civilization in the Amazon, going out alone at night seemed to be pushing it. But I couldn't help it. Nighttime in the jungle is both intoxicating and terrifying.

It took some time before my pounding heart began to relax and I became lost in the wonders around me: a brown leaf mantis waiting for prey, the metallic plumage of a roosting blue-crowned motmot, night monkeys. At night in the forest it is a battle of stealth. Foragers must find food in the void without making a sound that would give prowlers a lock on their position. Everyone strives for invisibility. JJ had explained that on some nights even a bright moon can cause the whole jungle to shut down: for creatures made to see in the most complete dark it's like being in broad daylight. On night walks we would always wait for dark, cloudy ones, when there was no light whatsoever. That's when the jungle came alive. This night was ink black and everything was moving.

How the Amazon gained the staggering multitude of species that live within it remains largely a mystery. We do know that the story began a few hundred million years ago, when the continents of our planet were jammed together in a landmass scientists have come to call Gondwana. Africa and South America were spooning and across them both flowed a tremendous proto–Congo River system that started in the east and flowed west across Africa and what is now South America to drain into the Pacific. When the continents began to drift apart in the late Jurassic the great river was broken in half, leaving Africa with only the upper reaches, which would later become known as the Congo River Basin.

Over the next 130 million years, South America drifted away from Africa, forming the Atlantic Ocean. The low-lying riverbed that had once filled from the Congo became filled with the salt water, until the continent jammed into the Nazca Plate, a geological giant that halted the drifting landmass and spiked the Andes Mountains out of the earth, blocking the westward-flowing river. For a few million years the entire Amazon Basin was nothing more than a massive inland sea; the stagnant, continent-sized swamp gradually desalinized and turned to freshwater. The slow transition is why many saltwater creatures adapted to freshwater. Today more than twenty species of freshwater stingray probe the riverbeds of Amazonia, along with pink river dolphins, manatees, and others.

When an ice age caused sea levels to drop, the Amazon swamp began draining into the Atlantic. The Amazon River was born. Such was the start of the river system as we know it today. In the hot equatorial climate and abundant moisture, jungle flourished. The conventional theory for what happened next has to do with the coming and going of ice ages. Scientists believe that as the world was clutched in periods of glaciation the cooler climate caused vast areas of jungle to die out and become grassland, while other areas survived as jungle oases. This created an isolation effect. Over tens of thousands of years the jungle would have survived in patches, along with the species contained therein.

The phenomenon of unique species living in isolated communities is called allopatric speciation. In the Amazon, it is believed that this happened again, and again. Each time an ice age came, the Amazon would turn to an archipelago of jungle islands, intersected by grasslands. Each time the thaw returned and the temperature rose, the flourishing jungle would rejoin the isolated groups. And so the Amazon went through stages of intensifying and then mixing. Even today grassland savannah makes up a large portion of the Amazonian tapestry.

But this theory of isolation causing speciation has its critics. Increasingly it is believed that interspecies competition was the largest driver of species proliferation, the challenge to adapt or become extinct. Whether the cause was climate or competition or, more likely, a combination of both, the result is the greatest explosion of life to have ever existed.

In my night walk I saw ants of various color and size racing along the ground, over vines, and up trees: leaf-cutters, army ants, solitary bullet ants, as well less familiar ant species symbiotically inhabiting and guarding pterocarpus and triplaris, cecropia and other plants. Jumping spiders, social spiders, ornate webs, simple webs, and hundreds of sets of octo-eyes caught the light of my headlamp. I could count more than seventeen different tree frog calls, and bats of every variety raced by, beating their fleshy wings and chattering. Well over a hundred different species inhabit the Madre de Dios, each with its own specialized niche: some fish, others eat insects, some drink nectar from flowers, while others suck blood from live animals.

Looping from the Brazil nut trail to Transect A, I entered a small stream with water up to my waist in search of my own prize: anaconda. The question of why there were apparently no anacondas in Las Piedras was puzzling. The only evidence of their presence had been at the lake where the farmer that shot Lulu's mom showed us the skeleton of the twenty-footer he had killed. Walking in the stream, I was hoping for a glimpse. They had to be here.

The hours that passed held many wonders, but, to my frustration, giant snakes were not among them. Instead I spent my time watching the subaquatic inhabitants of the stream. In one bend there were more than a dozen species of fish. Sucker fish clung to drowned tree limbs while another species with leglike feelers probed the aquatic environment. A slender eel-like fish with a pulsating lower fin running half the length of its body moved smoothly forward and backward. Larger catfish lurked in the deepest parts of the pool. There was an exquisitely plated hypoptopoma, and then a small spotted fish with a forest of vertical feelers sprouting from its face, some kind of Ancistrus. How could so many forms exist in such a tight space?

As I was contemplating this, something moved mere inches from my face. It was a huge black tarantula stalking down a fallen tree branch toward the water. I usually find tarantulas charming, but I was alone and already on high alert, thus the sight of the spider so close to my face gave me a jolt. This was the legendary bird-eating spider of the Amazon. The dinner-plate-sized hairy beast with inch-long fangs walked calmly down the pole beside me; when it reached the water's surface, it casually continued down the pole and disappeared into the pool. That was enough for me, and I made my exit from the stream.

Later that night I held the gaze of either a puma or a jaguar. I couldn't see the animal itself, only its lantern-green eyes searching to see what was behind the strange light. Emma said you had to earn a jaguar sighting. I also spent a spell in a tree when a herd of peccary cornered me. Despite moments of fear and even being lost for some time, the jungle night remained too alien and fascinating to deny. As first light rose in the east I realized that in one night I had been more alive and packed in more adventure than I had in my entire pre-Amazon life. It was fully light by the time I got back to the station, and I called out to Lulu to announce that I was back. But she was nowhere. I found a cigarette butt in a coffee cup, which meant Pedro had come by, and I figured he had taken her down to the farm with him. So I headed down there, too.

I found Pedro beside the river, hammering the blades of a propeller into shape. I approached and he looked up excitedly. "I saw a jaguar this morning," he reported in Spanish, "right over there," gesturing with the mallet.

"Really?" I asked, slightly perturbed that I had missed the sighting. Pedro responded that he had seen the predator at the station in the morning, out in the open. I looked around the sugarcane and banana trees expectantly. "Where's Lulu?" I asked. Pedro looked up, eyebrows raised. The expression on his face spoke clearer than any words could have: *you mean she's not with you?* I turned on my heels and sprinted the path through the forest and up the palm-plank staircase.

Racing along the last hundred yards of the path and emerging into the clearing of the station, I called for Lulu. I called repeatedly, frantically running around the station. It was too much of a coincidence. Lulu was nowhere to be found. But there were large, newly laid jaguar tracks behind the kitchen.

I cursed myself for leaving the anteater alone and retraced the entire station calling for Lulu. Once again the search produced nothing. After a night of no sleep, shell-shocked by worry, I wiped the sweat from my forehead and tried to think.

Pushing aside the burlap door, I entered my room and sat. Hands on my temples, I tried to think of why Lulu would not be at the station. Could she really have been eaten? I knew she could. After all, Emma and JJ had tried to start a chicken coop and the birds had lasted mere days, picked off by a legion of predators. If an adult jaguar had grabbed Lulu, there would be no evidence, no sounds; I'd simply never see her again. Panic was beginning to set in when there was a noise under my bed. The sudden commotion below startled the hell out of me and sent me springing to the other side of the room. Lulu's face emerged cautiously from under the bed, and then she lunged for me on her hind legs. I scooped up the little anteater and showered her in affection.

She had been hiding under the bed, something she had never done before. As I carried her out into the sunlight and toward the kitchen for a meal, it seemed certain that the jaguar's presence and the anteater's actions were related. I could tell she had been scared, and her excess energy in the wake of her fear was comical. She wanted to play.

She balanced on three legs, stretching one arm toward me, like she was calling me out, pointing. Her head was held down at an angle and she grunted and scratched at the deck, advancing and then falling back. I tried to pick her up but she slashed at me. I tried to grab her, but she slashed again. Dropping to my knees, I started a gentle slap-boxing session with her that she enjoyed immensely. My job was to get to her head; hers was to keep me from doing so. I haven't mentioned that through all of Lulu's affection and seemingly tame behavior, the one no-no was touching her head. She loved being cuddled, cradled, and even wrestled with, but if you touched her head she'd punish you for it. The behavior was likely born from the evolutionary advantage of keeping safe the delicate cartilage of their snout.

Another idiosyncrasy was her fondness for using other objects to scratch herself. Standing next to a tree or pillar on the deck, she would rub back and forth, careful to find the right spot—sometimes standing to get the hard-to-reach places on her back, like a bear. Or she would hold her head high and fully extend her ten-inch tongue until it could go no more, then shudder it violently and retract. It looked like an anteater's yawn. She would also spend hours sucking on my finger while I read. Completely edentulous—having no teeth whatsoever—the anteater would entirely devour my ring finger inside her long snout, tongue lapping within. I think she liked the salt.

Anteaters would most likely not make most people's list of intelligent animals. We are familiar with primates, canines, dolphins, and other high-order mammals as intelligent, but still underestimate what they are capable of. Tigers, of course, are members of the cat family, a commonly known super-intelligent animal group. Yet many people don't realize that they need to teach their young to hunt, and how to stalk, and find water, to survive in the wild. Without the guidance of its mother, a tiger cub would be dead in a matter of hours, and tigers raised in captivity can never adapt to conditions in the wild—they need the *knowledge* of their mother. This is a massive problem for conservationists as they try to rescue tigers from extinction; once the chain is broken it can never be repaired.

Another example of obvious imparted knowledge in the animal world is the mother otters that *teach* their young to dive to the ocean floor, and retrieve rocks so that they can float on their back and crack oysters on them. This is a complex skill that, without instruction, an otter would not discover out of instinct. What's more is that mother otters even show their young how to hold a good rock in their armpit while diving for more oysters, so that they don't have to waste energy searching for another tool.

The skills of many animals are often attributed to "instinct," which in the case of otters and rocks, or tigers and their cubs, is just counterfactual. The reality is we still know very little about what animals think or how they feel and communicate. Lulu was an animal that showed an incredible range of feelings, from fear to unbridled happiness. She was affectionate and attuned to my actions in a way similar to a dog (which is saying a lot). A wild animal, profoundly bonded to a human.

Lulu depended on me emotionally and physically for survival, and I was determined to do everything I could to protect her. Even as a young naturalist I knew that the cards were stacked against her. I could never teach her the things her mother would have. It was doubtful that she would be able to survive on her own, but I had hope. At least at the research station and surrounding preserve she would be safe from hunters. She'd also have access to a vibrant ecosystem that had not been compromised by the guns and chain saws that had pillaged the rest of Las Piedras. Here JJ and Emma's work over the last decade had protected the animals and the ecosystem. She would be safe from my species, her greatest threat. Each time I'd feed her I would fit that old spark-plug protector-nipple onto the bottle. And each time I felt a fierce warmth at the thought that a piece from a dismantled chain saw, the object most lethal to forests, was being used to nurture the young life.

As our two weeks together ended I made a decision. It was a bright sunny day when Lulu and I were walking on a trail on the opposite side of the river. I looked around me and realized that there was no way I could leave. Damn the semester that would be starting in a week—how could I part with my anteater? This was a once-in-a-lifetime event. To anyone else it might seem pedestrian, but at nineteen I was still learning to take control of my world. Blowing off a semester was a big deal. As I walked, talking gently to Lulu, daydreams filled my head. I imagined what staying in the jungle for months on end would be like; I imagined watching Lulu grow. It would also mean that JJ and I would have a lot more time together, which I was eager for.

I had walked a small distance ahead of Lulu, who had stopped to sniff in the foliage. Turning to wait for her, I observed as she was surprised by a large yellow-footed tortoise that emerged from the brush beside her. The massive domed carapace of the reptile stood easily a foot from the ground and surely measured double in length, by far the largest of its species I had ever seen.

The huge tortoise startled Lulu, and, acting in defense, the anteater exhibited the signature retaliation of her species. Rearing up on her hind legs and spreading her hooked arms, she slashed at the tortoise, releasing a long and powerful trumpet sound. The unsuspecting and ancient tortoise rapidly altered course and retreated as Lulu laid siege to its shell. Again and again she whacked him, and the turtle started running for its life. Lulu was not satisfied and gave pursuit, smashing at the thick shell with surprising ferocity.

Given their peculiar anatomies, neither animal was in any real danger, and both emerged from the encounter un-scathed. Lulu remained standing for a short time after the tortoise's retreat, still grunting and trumpeting with arms outstretched and chest puffed out, like a teenager who has just won a fight spreads his arms and dares any other challengers to step forward. Replacing her trumpet sound with an aggravated "who wants some!" would have fit the image to perfection.

After determining that the situation no longer called for aggression, Lulu dropped back down onto all fours and came trotting toward me, flaunting her new swagger about the jungle. It was a spectacular day, spent in the sun and warmth of the Amazon. After returning across the river that night, Pedro and I cooked dinner and enjoyed a wild knife-throwing competition on the moving targets of the large jungle cockroaches that scurried across the kitchen wall. The night ended in numerous cockroach fatalities and much laughter.

But the next day something strange happened. Pedro and I were walking along a stream in the afternoon, fishing, when I threw up and fainted. There was no warning. One second I was fine; the next everything went black and white and my legs collapsed. Later that night I felt horrible and weak in the knees, and retired early to my bed instead of a hammock. Before sleeping I noticed that my arm had several pus-filled bubbles on it, and that my face, where I had shaved earlier in the day, was throbbing.

That night my dreams were a deranged and torturous

eternity. I was lost in the blackness of the jungle. This was no ordinary nightmare. The sensation of terror was so grave I wondered how I could survive. I prayed for light, begging my mind to wake, tossing and turning in my bed, while the force of some inner tide dragged me toward darkness.

7
The Descent

Roosevelt was in grave danger. The skin around his wound had become red, swollen, hot, and hard, and a deep, pus-filled abscess had formed on the soft inner portion of his lower thigh. . . .
—CANDICE MILLARD, DESCRIBING FORMER PRESIDENT THEODORE
ROOSEVELT IN *THE RIVER OF DOUBT*

Something was horribly wrong. I was awake but couldn't open my eyes. I tried to sit up, but pulling my face from the pillow caused pain so devastating that tears welled in my closed eyes. I blindly felt my way to the bathroom, where I used water to emulsify the powerful crust that held my face to the fabric of the pillow. Nearly fifteen minutes passed before I was free. As I laid the pillow down I saw that it was covered in the red and yellow streaks of pus and blood. In grim curiosity I lurched toward the bathroom to the sink and stepped back a few paces to where the morning sun illuminated my face in the mirror.

What I saw was a face indeed streaked in flowing blood and festering yellow-green pus. Deep recesses in the skin had been hollowed out during the night and now changed the landscape of my face completely. Where I had shaved above my lip the night before, a field of numerous fluorescent bubbles was now growing. Several mosquito bites had become volcanoes of pus as well. And it wasn't just my face; much of my body was covered in the painful, red, yellow, and green bull's-eyes.

The following days are a feverish haze in memory. I stayed in bed and could barely eat. With a temperature peaking at over 104.5 at times, I thought I was dying. I was living in abject fear that I would never see my family again. The illness poisoned my dreams, turning them into horrible nightmares that lingered even in consciousness. Suddenly the remoteness of Las Piedras felt altered; jungle stretching for unbroken miles in every direction had turned from freedom to grim finality.

Pedro appeared after two days and was shocked by my condition. The moment he saw my face he set up camp at the station. It scared me that he was trying to hide his emotions, and would sneak glances at me, in disbelief at how ravaged my skin had become. He cared for me as best he could and provided numerous jungle remedies. Nothing worked. We listened for boats on the river, my only way to help, but heard nothing for days.

The thought of leaving my anteater made my throat tighten, but it seemed inevitable. She was restless and frustrated that I wouldn't allow her to lie on me or lick my face, but I was adamant for fear that she could catch the infection. I tried making her sleep in a spare room by herself but she hated it.

During my last night at the station she came to find me, poking her head under the mosquito net and purring. I woke and walked to the bathroom and Lulu followed, bumping tiredly along the deck in the moonlit night, just wanting to stay close. Together we returned to the main deck and then to the room, where her sleeping box lay. With front paws on the

rim of the wooden side, she turned and delivered an affectionate nuzzle, which I accepted despite the pain. Climbing inside
the box she paced tight circles, as was her custom, continuing
around for several laps before falling into position with her tail
over her nose. Scratching her belly and talking gently, I let her
nibble a finger and hug my arm with her claws, before tucking
her in with her own bushy tail. It was the last time I saw her.

When the faint sound of an approaching boat finally did
come it was at 5:20 A.M. Pedro sprinted down to the river
and begged them to take a sick passenger. After they agreed
he sprinted back up to the station to grab my pack and disappeared, shouting over his shoulder on the way back down to
the boat that it was now or never.

I stumbled through the chilly morning, making for Lulu's
room to wake her and say goodbye. But she wasn't there. No
amount of hoping, praying, or pleading changed that fact.
Heartbroken and bitterly crushed, I made my way down to the
river and piled onto the boat. Pedro smiled as he waved me off,
and I tried to smile back.

Once again I was traveling alone on the Las Piedras, but
this time I was not surrounded by a family. Barely able to lift
my head, I looked around to see the unflinching scowls of the
roughest men I had ever seen. One was emaciated, one fat, one
had a thick mustache like a black squirrel's tail, one was old
with prickly white stubble, and another was young, maybe in
his mid-twenties. They were dressed in camouflage from head
to foot and armed to the teeth; it took me a moment to realize
what about them made my insides churn.

As we sped with the current I looked around. Piled under
the bow of the boat were seven or eight massive yellow-footed
tortoises, like the one Lulu had battled, bound in their shells

by balsa bark restraints. To my left was a blanket roll that encased the feathers of several freshly killed macaws. The rolled skin and decapitated head of a monster black caiman baked in the sun amid a cloud of black flies. In a shadowed box a live infant macaw was captive. The men had cut down the tree, killed the parents, and taken the chick. Two live pale-winged trumpeters were fastened to a beam, and in the shadows of the bow tied by its ankle was a baby spider monkey. Around the boat were bags of Brazil nuts, buckets of butchered meat, bags of charcoal, and other loot unidentifiable from my position. They were poachers.

Throughout the world there are various kinds of poachers, ranging from the poor farmer who sneaks into a protected area trying to nab some extra meat for the family pot, up to the highly organized and well-funded war criminals who shoot elephants with rocket launchers. These guys were a grade below the baddest, but they were definitely professionals.

The sun crucified my wrecked skin in the open-topped canoe. Winding through the jungle, we sped down the rapid waters of the swollen river. During this time I drifted in and out of consciousness. At one point, moving to the bow, I offered water to the baby spider monkey that was taking refuge in the shade below. With wide, jet-black eyes the tiny primate denied the offer, instead grabbing my finger and making desperate distress calls. Its mother lay crumpled and charred, stuffed into a pot in the front of the boat.

I knew that what these guys were doing was threatening the entire Madre de Dios. At one point the oldest of the men unwrapped the body of a blue-and-yellow macaw, which he held by the joint of the wings, glistening blue jade in the sun. Even in death the bird was so massive and flawlessly brilliant that it barely looked real. It wouldn't take very many crews like this one to wipe out macaws in the Madre de Dios; in fact the birds are an easy shot when feeding on the riverside clay licks. If the macaws disappeared the tourism industry would be hurt, and as a result employment would go down. Without jobs in conservation, many locals would return to extraction of resources like gold, timber, and wildlife. It's a simplistic example but illustrates how effects can snowball and repeat. This was the front lines of the extinction crisis.

I had grown up seeing photos and reading stories about the elephants in Africa massacred for their tusks, or rhinos blown away for their horns; tigers being exterminated for their bones. I had read of the North American bison and how before the late 1800s it had dominated the continent, in numbers that have been estimated as high as a billion. In each case there were the hunters and the buyers. Elephant tusks make luxury ornaments; rhino horns make fake medicine, as do tiger parts. Macaws make exotic pets; their feathers make exquisite decorations. Mahogany makes nice furniture. If people who bought these things oceans away only knew what they were funding . . .

These and so many other stories were the stuff of my childhood nightmares. It rattled me to the core to imagine a world without the creatures that make it colorful. Now, on board with the poachers, I was living the nightmare. Around me the species I had tracked in the forest, photographed, and wondered over with such appreciation were cruelly butchered.

We had traveled for three hours when one of the party said he smelled peccary, and the motorista cut the engine. We drifted to shore, where a deluge of insects began attacking both the dead wildlife aboard the boat and my razed skin. I would have remained on the boat while the men disembarked in search of the pigs but the bugs were too bad. They were laying siege to my open sores, and the pain combined with the visual of pus-filled craters swarming with feasting flies made me even sicker.

Feebly hobbling behind them as they energetically stalked the herd, I felt my brain once again preparing to faint; the fever was unbearable. I don't remember much from the hunt. Through the fever my mind recorded events much like it does when blackout drunk, in brief images. It must not have taken us long to find the herd, but I remember there being boar everywhere around us, perhaps seventy of them. I remember two guns roaring in tandem, transforming the jungle into a war zone. The herd of black pigs erupted in frantic terror and stampeded in every direction. Deaf except for the loud ringing of ears, I watched as two fell to the ground. Again the guns fired. More pigs fell.

One confused boar ran toward us and fell. Peppered across his face and chest were numerous impact points where the buckshot had made contact. He fell to one side, legs running in the air as bodies everywhere writhed on the ground. Another peccary, a female, ran into the forest screaming as blood dripped from her face: the pellets had broken her skull but not enough to kill her. A piglet searched desperately for its mother nearby. The man with white stubble kicked the orphan with all his strength, sending it through the air and into a tree. In the wake of chaos came a terrible silence. The sound of the large wounded boar dying filled the air.

Stepping over one of the injured pigs, the gnarled-looking motorista gave a rude poke to the animal's stomach with his machete. It was still very much alive, gasping through the holes in its lungs, suffocating slowly. Shifting position and using the blunt edge of the machete's blade, he tried to break the animal's skull. *Crack!* Each time, the blade hit the impassibly thick crown of the skull. As the blows fell, skin and blood began to fly and the impacts echoed against the dismal silence of the forest. My hand went to my forehead; the thought of what those blows must feel like brought tears to my eyes. I prayed that the hog had gone into shock long ago and wouldn't feel its violent death.

While the other men were skinning the pigs, the youngest one slung a gun over my shoulder and told me to follow. He wanted to go fishing. Beside a small lake, he shot a monkey for bait. The primate struggled in the vine as the life drained from her body; the poacher sat beneath a tree to roll a cigarette. Unable to control my rage, I spoke. "Esta viva—she is still alive!" I spat at him. But he only shrugged.

Pulling the .22 from my own shoulder, I checked to see that it was loaded, drew a round into the chamber, and took aim. Fixing the sights on the monkey's chin, I exhaled before pulling the trigger. The shot traveled upward through the center of the tiny horseshoe-shaped lower jaw of the monkey's skull and out the top of the cranium in a spray of pink. She fell to the ground. I had never shot anything before.

By shooting the monkey I had overstepped my welcome, and from then on the poachers grew hostile toward me. Thirty minutes later we were back on the boat with four butchered pigs and two dozen piranhas. The remainder of the day stretched on into eternity while I continued to fade in and out. Most of it was unremarkable, save for when I was awakened by gunfire. Throwing aside the tarp I was under, I rose to see the men staring at an empty beach. They explained that they had shot at a large anaconda. Thankfully, they missed. Regardless of the circumstances, my heartbeat hastened in wonder: *an ana-conda*. My mind was too weak to dwell on it, though. Instead I descended into a deranged state, an endless ticker of random images and sound bites.

As the hours wore on we covered miles of clay-colored river. Surrounding us as we traveled were the massive carcasses of fallen trees. Flooded from the storm the night before, the river had risen almost ten feet overnight, gaining force and wearing away the banks on many bends. The eroded clay had released trees hundreds of years old into the river. Our motorista took great care navigating the churning waters and swarming mine-fields of debris. Once when a tree trunk four feet in diameter struck the ground, the current lifted the back end out of the water; it towered forty feet above our boat before crashing back downward with chilling force, narrowly missing us and soaking everything on board. At another rapid bend in the river our boat was viciously broadsided by the current of a massive whirlpool more than seventy feet across that threw some of the supplies on board over the side. The swirling current held our vessel as if it were a blade of grass for several rotations amid bus-sized chunks of timber that shook us violently. Ev-eryone was alert during these moments; even I managed to lift my head for a time. The weight of the trees around us could have turned the small boat to splinters if we got pinched. It took the tightening of every sinew in the boat driver's forearms and the gravest furrowing of his brow; waiting, maneuvering, and then waiting some more until a window appeared, he then gunned the engine. The dangerous exit sent us rocketing free once again downriver.

Back in Puerto, I made emergency travel arrangements to fly out the following day and spent the night in a hotel room,

swathed in Vaseline, never happier to be watching *Seinfeld* re-
runs. In all it was seven days between the moment I realized
I was in trouble and my arrival home to JFK Airport in New
York. On the flight home I had three separate plane connections to make, and each time I found myself with an entire row
to myself. Several different people in the course of my numerous return flights discreetly got up from their seat after seeing
me and never returned, undoubtedly asking the stewardess for
another seat. Walking to my gate before flights with a swollen
and disfigured face, covered in highlighter-green ooze, I was
an obvious leper and drew the stares of hundreds. I still can't
believe that they let me fly.

At JFK I experienced the usual change-of-worlds shock,
which was returned by the customs officer who checked my
passport. Looking up to see if my face matched the one on my
passport, his eyes popped. "Buddy, what happened to you?"
he asked in a blunt New York brogue. I smiled, though it hurt.
"That's why I'm home, man. I need a doctor!"

"Jesus Christ! Well then, welcome home, dude," he said, shaking his head as he stamped my passport, "Now go! Go! Go!"

At the emergency room doctors took one look at me, heard
the word *Amazon*, and got me into an air-locked room where
doctors entered and left wearing hazmat suits. Yet when tests
came back they found it was not any sort of disease. What I
did have was a horrible MRSA (methicillin-resistant-staphylo-
coccus aurens) infection of the skin. The infection had spread
over all the open wounds, from mosquito bites, to my shaved
face, and to my eczema, and had devastated the entire epi-
dermis. I spent four days in the hospital on IV support. The
doctors repeatedly told me that had I not made it back when
I did, I might not have made it at all. But it would not be my

last run-in with illness in the jungle; far from it. In the years to come I would endure botflies, tularemia, numerous pique parasites, dengue, and wicked relapses of MRSA.

All told I came to spring semester three weeks late that year, though it took some explaining and finagling. Telling the story of why I was late to professors sparked some lasting friendships.

To this day I am heartbroken to have left Lulu. Given the choice, I wouldn't have missed a moment with my anteater, and if I ever find myself with a time machine and some powerful antibiotics, I know exactly where I am headed. Emma and JJ continued caring for her after I had gone, and as weeks went by, her solo excursions in the forest become longer and longer. She gradually got the hang of eating ants, and eventually returned to the jungle, leaving us to wonder about her fate.

That she came to live entirely in the forest is a major success. Given her tenacity, I am confident that she'd be able to fend off jaguars, and with the rich forest around the station she'd have an ample supply of food. Though I missed out on her last days at the station, all I could do for her was to continue to work with JJ and Emma to protect Las Piedras.

I continued to bring and lead volunteer expeditions, and my skills further improved. My relationship with JJ kept growing and we spent more time before and after volunteer trips, having adventures of our own. We discovered that without others to look after, we could travel harder, faster, lighter, and more quietly to elevate the caliber of our explorations. But we still weren't finding anacondas. JJ began asking around, talking to his family, and quickly arrived in the counsel of the man with the answers: his father, Don Santiago.

8

The Wild Gang

We need the tonic of wildness. . . . At the same time that we are earnest to explore and learn all things, we require that all things be mysterious and unexplorable, that land and sea be indefinitely wild, unsurveyed and unfathomed by us because unfathomable.
—HENRY DAVID THOREAU, WALDEN: OR, LIFE IN THE WOODS

Winter had descended on the jungle. Low clouds cast a somber light on Puerto Maldonado. The locals call the phenomena *friaje*, a seasonal period when cold mountain air from Patagonia travels northward along the Andes, gripping the lowlands in a temporary chill that drops the temperature as low as sixty degrees. As a result the usually steamy jungle town appeared silent and unfamiliar. Instead of being bustling and tropical the streets were quiet, and the few people walking about were insulated from head to toe in what northerners would recognize as snow gear. Children were bundled like Eskimos by protective parents: hats, gloves, coats, and scarves. It was a comical overreaction to weather when a sweater would have sufficed. Nonetheless, I had no sweater with me and bore the motorcycle cab ride with clenched teeth. Arriving at Barrio

Nuevo, I dismounted the Suzuki dirt bike, and the driver extended a gloved hand, into which I placed two soles.

The previous night I had slept in a hotel room, after two weeks of working with Emma leading an expedition. Nearly a year after I had come downriver and parted ways with Lulu, having completed two more semesters at college, I was back. So far I hadn't seen much of JJ, who had stayed in town to carry on the legal battle and to take care of Joseph, who was beginning preschool. But on returning from the expedition, I had been told that JJ had made progress on the anaconda front, and he wanted to meet early the next morning to discuss it.

I checked JJ and Emma's house, which was vacant, then made for the Durand family residence farther up the road, a cluster of dwellings that sat atop red-clay cliffs, two hundred feet above the Tambopata River. As I approached the house I looked out over the confluence, where the mouth of the Tambopata pours into the Madre de Dios River's sweeping bulk.

I passed the rope-hinged gate into the open yard. There was no one in sight. There was the feeling of winter here, too. All seemed still—even the chickens were quiet. I rapped once on a flimsy door and was told to enter by a voice I didn't recognize. I entered a room made from boards, glass, steel, and tarp. The interior was without light, and bodies crowded around a table in the dim space.

When I entered, the Ese-Eja family sprang to life with greetings. The table was a war zone of beer bottles, cigarette butts, and playing cards—evidence of a productive morning. Exchanging hugs with Chito, Robin, Pico, Melissa, and many others I made my way around the suddenly merry room of drunken people. Pico seemed to be the farthest gone. He bellowed my name and wrapped a powerful hand around my wrist, dragging me into a rough bear hug.

During the last month leading volunteer expeditions at Las Piedras, Pico had come as the designated motorista. Over the course of those weeks Pico and I had formed an uncanny bond. He was entertained by my love of the jungle and would burst into uncontrollable laughter, swearing and clapping his hands when I jumped onto a caiman or caught a snake. He thought I was crazy. Despite the language barrier, we communicated well and spent hours yakking a profane interlingual gibberish and having a great time. The day before, he had piloted the entire return journey from Las Piedras station to Puerto. Now he was enjoying his first day off in weeks in exactly the way he liked best: wasted.

He forced me to sit, sending his daughter Kiara to fetch her grandmother. I tried scolding him and the others for drinking in the morning, but they would hear none of it. Apparently this was the only way to spend a cold friaje Sunday morning. Pico was talking a hundred miles per minute as he fit my hand into that of Robin, who lifted his head from the table where he had been drowsing. Robin was the *muy guapo* of the brothers, the charming football player who got all the girls. Women were defenseless against his big, sultry eyes, careless but perfect hair, lazy swagger, and tortured Latin ooze. At the time he had already earned the reputation of ladies' man, even before he became a licensed wildlife guide with a big job at a tourist lodge on the Tambopata. And there he enjoyed a constant rotation of international *chicas*.

Robin and I spontaneously arm-wrestled with locked fists. Much noise resulted from our long stalemate, and a powerful crescendo overtook the room when he slowly but surely brought my arm to rest on the table. We each downed a glass of beer. JJ's mom, Doña Carmen, entered the room next, pushing aside the cigarette smoke and bending over to give me a rough kiss on the cheek and throw an alpaca sweater over my shoulders. "Look at my crazy boys!" she boomed in Spanish, as if ashamed of their state. She slapped Pico on the head and shot me a wink. There was nothing she loved more than having a crew of her children at home and under her roof. Over the din she shouted, "Have you found JJ yet?" I told her I had not, and she rolled her eyes.

As Robin and I prepared for our second battle, this time with our left arms, Doña Carmen swiped one of the many phones scattered atop the table's wreckage and called JJ to tell him I had arrived. With one hand pressing the phone to her ear, she used the other to sling the dead duck she had been holding over her shoulder. Matriarch of the family, she had treated me as one of her own from the day she met me. In her mid-seventies, she was more Viking than old mare. She almost always was carrying an ax or machete, ready to disembowel or chop up whatever her sons brought home for her pot. Today it was duck for lunch, and I could see piles of spiny bones on the dirt floor. I had missed a catfish breakfast by only a few hours. She asked me if I wanted to eat and I told her I already had. Moments later she slipped a piping-hot bowl of catfish soup under my nose. It was a delicious construction of jungle ingredients: catfish, rice, herbs, and water. Her husband, Don Santiago, I was told, had caught the fish the day before.

I heard the purr of a Yamaha motorcycle pulling up to the house, and a moment later JJ entered with his signature gusto and energy, removing the helmet from his head and clutching me in a powerful hug. I laughed; we had just seen each other yesterday! But today was different and his eyes were wild. He pulled a chair backward up to the table, as men do when they have something inordinately important to say. Leaning forward with intensity, he placed a hand on my arm, and with the neutral cadence of his broken English said: "We go to La Torre." It was half statement, half question, and it silenced the room.

My eyes widened, and more than one brother perked up at the mention of the river. Pico, who had fallen to the floor laughing moments before, suddenly demanded to be lifted back to his seat, and even Doña Carmen pulled up a chair. "Verdad?" I asked. *For real?*

"Sí!" said JJ. The conversation continued in Spanish. "We must go, the cold is the time. I spoke to my dad yesterday. We will find many anacondas." Pico slammed both fists on the table in excitement and bellowed; a map was passed over my shoulder and spread on the table. JJ continued. "With the cold, the snakes will come out of the water looking for sun. This is the right time to go!" Doña Carmen nodded her head knowingly and was the first to speak, asking JJ whom he intended to take. Thinking for a moment he used his fingers to list himself, Pico, Chito, and me. Chito looked nervous.

We planned as if preparing for a dangerous heist. Each brother offered advice on locations and strategies for finding anacondas, citing decades of expedition knowledge and various locations on the map—none had ever gone out with the intention of finding snakes before. Paper was found, and we listed the supplies needed. Coffee was brought in, and our crowd leaned tightly over the table. From the tone of the conversation the weather seemed to guarantee that we would find anacondas. I saw the family as I had never seen them before.

More than one brother expressed relief at not having been selected for the excursion, labeling our party loco for intentionally seeking the great snakes. Pico shot me wild smiles of anticipation.

As we ate and talked, Doña Carmen stood over us shaking or nodding her head, depending on whether she approved or disapproved of the conversation's content. She took pride in her family's thorough planning. From across the table I saw her watching me out of the corner of her eye. First a drunken Sunday and now an expedition into ancestral hunting grounds restricted to all but the Ese-Eja; I was being drafted into intimate corners of her family's life and for some reason, she approved.

Elías emerged from the opposite room and slapped a four-fingered hand on my back and smiled. "Cuidado tus huevos," he said. *Watch your balls.* As he smiled, the seven-inch scar that traveled from his chin to his temple twisted. Elías was one of the elder brothers, in his early forties. I barely knew him at the time but would later learn that he was the toughest of the crew and had the reputation for ending fistfights with his head. I told the man missing a finger and half his face that I'd be careful. He gave me a "rock on!" smile, lit a cigarette, and nodded to the music.

Several times I was asked if I knew what I was doing; no one they had ever met had wanted to *catch* anacondas. I told them that I had worked with snakes my whole life, and JJ vouched for my skill; he had seen it. But now JJ's older brother José spoke up. "Juan, I don't think you understand what you are doing, and neither does he. Do you think this is a joke?" José tilted his head in my direction. "You can be crushed in seconds; people have been eaten."

"Oh, come on!" I said. There was no way that was true.

"It's true! His father-in-law was swallowed whole!" José yelled with his open palm pointed toward Elías. We all looked to Elías, who was now rocking in a hammock, and he nodded: it was true. In later years I would learn that the man had gone to check his boats by the riverbank one night, and never came back—in the morning the largest anaconda anyone around had seen was observed just next to the boats, with a human-sized lump in its gut.

The thought of man-eating anacondas was making Chito's face turn pale, and he was almost completely silent during the entire conversation. Almost all the brothers were experts in all things jungle, and had learned from their father and worked as farmers, loggers, or hunters at some point in the past, if not their entire lives, but Chito, barely into his twenties, had not grown up in the forest as the rest had. He was terrified of snakes. I would find out later that JJ and Pico had selected him deliberately because of his shy nature. As we talked about the dangers of anacondas, he grew still paler. I told everyone that I had it covered when it came to snakes. They approved my participation just on the basis of my confidence alone. And so the expedition was set. Tomorrow we would head into the jungle as *anaconderos*: anaconda men.

The creation of Bahuaja-Sonene National Park is a landmark in the history of wilderness stewardship. Its story begins in the 1980s, when American veterinarian Max Gunther purchased a 260-acre hunting reserve at the confluence of the Tambopata and La Torre Rivers, and converted the land into a tourist destination. Gunther's lodge, which was named Explorer's Inn, was a tiny spot, a teardrop on the beach compared to the ocean of pristine jungle behind it.

In the beginning, Explorer's Inn was a small operation and attracted few visitors. The wildlife had largely been scared away or killed. But still, though modest in number, people were drawn to the area. This fascinated a young Ph.D. graduate from Princeton named Charles Munn.

Charles Munn had already unlocked the secrets of macaw reproduction while working in the Manú region, making him a legend in the field of conservation. After finishing his PhD at Princeton, Munn approached the Wildlife Conservation Society (WCS) for a job, but explained that he did not need the money. He warned his boss that if the society did not let him do what he wanted, he would leave "in a heartbeat." It was during this time that Munn, fascinated with Explorer's Inn and the Tambopata, began devising an audacious goal: to protect an entire ecosystem. Munn and other scientists devised a radically aggressive plan to save the region in its pristine state, an objective never before achieved. For more than a decade Munn pursued a plan that others considered insane: to designate the entire Tambopata and Candamo river basins as one protected national park.

Traditionally, protected areas must exist between inhabited areas and cannot encompass an entire ecosystem. This policy has been followed across the globe, but Munn saw a unique opportunity to shatter the mold. Why should the fact that no one lived in these areas preclude them from protection? But to protect the drainage of both rivers, an area between one and two million hectares had to be set aside. The audacity of the plan was laughable, yet Munn persisted. These areas were largely uninhabited and would never again be such prime candidates for protection. It was now or never.

Unlike most famous conservationists, who promote them-

selves in the interest of furthering their work or who write books
to inspire others, Munn is virtually impossible to research. He
has been described as the conservation world's Bruce Wayne.
Little is publicly known about Munn except that he comes from
a wealthy family and has done things that no one else could.

Munn rallied the support of local people and indigenous organizations. He spent years cataloging the biological and ethnographical nuances of the proposed protected zone, showing
how they would benefit the Peruvian economy. Munn knew
that creating a park was not enough; he had to find a way for
it to make sense in the minds of Peruvian officials who viewed
the jungle as an economic opportunity, and he saw ecotourism
as the answer. The park's creation was linked directly to the
idea that local people would benefit from the intact ecosystem
through tourism. Along the way he received many threats and
was once nearly imprisoned by a corrupt forest minister. He
was doing so much so fast that the people who saw the forest
as a big dollar sign saw Munn as a menace. Years later he'd flee
Peru to escape threats to his life.

Munn remained undeterred while creating his dream park.
After he and hundreds of community leaders and conservation
experts spent years cataloging biological and ethnographical
data and completing exhaustive economic studies, and after
uncountable hours of local, regional, and national meetings,
the government approved the mega-park. Simultaneously,
however, the Peruvian government granted an oil-exploration
concession to Exxon Mobil smack in the middle of the proposed sanctuary. With this development came Peruvian president Alberto Fujimori's decision to cut the park's proposed
area by almost 50 percent.

The land granted to Exxon contained almost the entire

Candamo River. This was the most pristine watercourse of them all. The crown jewel at the center of the plan, the most important part, was suddenly in the hands of an oil giant. Along with the guaranteed roads and human activity that would follow was the looming threat of an oil spill; a leak in the Candamo would destroy the valley and leach oil throughout the entire rest of the Amazon.

In response to this tragedy waiting to happen, Argentinian conservationist Daniel Winitzky put together a team and began filming. The resulting documentary received overwhelming public support, and suddenly journalists from all over the world flocked to the planet's last Eden. When *Candamo, la ultima selva sin hombres*, or *The Last Forest Without Man* aired in Peru and abroad, the film ignited a frenzy of public interest. With locals, scientists, journalists, and the citizenry of Peru up in arms, everyone was watching for what happened next.

In the end Exxon Mobil must not have found substantial oil within the valley, because they backed down. On September 5, 2000, President Fujimori responded to the public storm by doubling the size of the park. The new protected area covered 3.7 million acres, or five thousand square miles, of rainforest. The area is located between the Tambopata and Heath Rivers, which in the native dialect of the region were respectively the Bahuaja and Sonene. Hence, Bahuaja-Sonene National Park came into existence.

In celebration, Munn wrote, "I have just finished checking with the world's experts on tropical forests and have concluded that without doubt the watershed of the Candamo River and the immediately surrounding areas of tropical forest in the Tambopata-Candamo Reserved Zone are the largest completely uninhabited, un-hunted tropical habitats on Earth."

9
Anaconderos

Going up that river was like travelling back to the earliest beginnings of the world, when vegetation rioted on the earth and the big trees were kings.

—Joseph Conrad, *Heart of Darkness*

The current pushed our canoe like a leaf as we dodged boulders and whirlpools, bending tightly upriver. The La Torre was a smaller river than Las Piedras. It was often no more than fifty feet wide, with long beaches on the shallow inside of each bend. Framing the beaches were tall stands of bamboo and river cane, behind which the jungle rose uncompromisingly large, decorated in places by flower tapestries that hung from the canopy. The intimate tunnel of green was alive with basking caiman and lapwings on the banks, jabirus, sunbitterns, and horned screamers. Just in the first hour we passed a pair of gold-striped tegus, lizards more than three feet in length; a large weasel-like tayra devouring a fish; and a harpy eagle that cruised over the river like a Harrier jet.

Pico leaned back, piloting the canoe in a tight sweep, and as we came around, a flock of blue-and-yellow macaws flamed out from the canopy into the air above us. Their emergence rained a cascade of leaves onto us, as over seventy of the massive birds flaunted plumage of radiant gold below and iridescent tanzanite blue above. We were engulfed in a vortex of color and sound that was difficult to accept as real. We all cheered and whistled along to the din of the magnificent birds. As they flew off I wondered what dullard had decided to name them "blue-and-yellow macaws." What an injustice to beauty; they were breathtaking. It was the first time I had seen them in real life, and at the sight of such a wondrous creature, the name blue-and-yellow macaws seemed almost offensively boring. As we chugged upriver I decided that using the Spanish word for gold, *oro*, joined with cerulean, would make a far better name for these otherworldly birds: oro-ceruleans.

Animals do not occur homogeneously throughout their range. Oro-ceruleans (or blue-and-yellow macaws) might be shown in a field guide to inhabit "the Madre de Dios of Peru," but they are not dispersed evenly throughout this range. Instead, they exist in communities. This is why they are found on the La Torre and not on the Las Piedras. Similarly, spider monkeys, abundant on Las Piedras, are absent on most of the La Torre. Animals require a breeding community of which they can be a part. Even though each river in the Amazon region may appear more or less the same to an observer, the species of plants and animals found can differ greatly. It was this feature of community that had prompted JJ to suggest we travel on the La Torre. This was anaconda land.

Often overshadowed by species like jaguars and giant otters in the ecologist's imagination, anacondas affect virtually every level of the vertebrate food chain. They start off small, eating prey like frogs, birds, fish, and baby caiman. As neonates, they themselves are often prey to these same creatures. But then they grow. At mid-size—eight feet for males and twelve to fourteen for females—an anaconda gains access to a wide variety of mammals, ranging from juvenile peccary to rabbits, birds, capybara, caiman, and armadillo. Once fully grown, these snakes prey upon capybara, peccary, tapir, caiman, and even the occasional jaguar. This species is an important riparian predator that influences the behavior of dozens of other species. Their impact on the ecosystem is far-reaching. The population of anacondas determines the population of capybara, which consequently corresponds to the variety and amount of vegetation by the river's edge, which in turn has implications for species of bird, mammal, and reptile. If you trace the thread of anacondas in the food web in the direction of caiman, now there are two predators whose lives are braided together, consequently enacting myriad influences on a number of prey species.

JJ and I sat in the middle of the boat, atop the supplies and food. Sweeping the landscape with his keen eyes, the Peruvian explained that anacondas would be found in the piles of timber lining bends in the river. *Palisadas*, as he called them, were the mountains of debris that the river harvested from the jungle; great trees were uprooted, carried on the current, and eventually grounded on the shallow bottom. Some piles stood thirty feet above the water, extending more than fifty feet from end to end. These palisadas were tangled masses of thorns, broken limbs, holes, and slick surfaces. I wondered how I would fare when hopping onto one of these piles, especially when in pursuit of a large snake. It didn't take long to find out.

JJ's arm shot forward, index finger extended: "There is! Anaconda!" I was amazed. *Really? Already?* Chito sprang into a ready stance and Pico positioned the boat. It took a moment for my eyes to light upon the snake, but when they did, I saw an anaconda larger than any snake I had ever seen. It was vividly olive green with large black splotches running its length. The creature was a very healthy individual about twelve feet long.

JJ and I perched in the bow, ready for when the boat would touch shore. We landed and gently disembarked onto the bank. I directed JJ to follow behind me. Thanks to Pico's deft boat-manship, we found ourselves beside the largest snake either of us had ever attempted to catch. Coiled among the two-foot-tall grass on a sloping bank of the river, the snake seemed to be asleep. I approached with caution, ready for either a defensive attack or an unannounced bolt by the snake for the river.

The snake chose the latter, and when we were no more than three feet from his perch, it lunged toward the water. I leapt full-out, diving onto the snake's body, joining its race toward the river. Together we slid splashing into the water. As the snake struggled to escape I gently snatched it behind the neck as JJ lifted the heavy coils.

The constrictor spellbound the three brothers and me, and despite a lifetime living in Amazonia, none of them had ever been so close to a live anaconda. Once on the beach we mea-sured him: eleven feet, six inches. Then he was placed into a large bag and weighed. We took note of our findings as well as several photographs before returning the snake to the wa-ter. In all we troubled the snake for only five minutes before departing.

We traveled farther upriver so as not to cause any further disturbance, and JJ and I plunged into the cool water to bathe, washing the mud and snake feces from our bodies. I remember tackling JJ in celebration. *An anaconda finally!* We all relived the capture and laughed for a time, but not long afterward we were back on the boat, winding deeper into the heart of Bahuaja-Sonene.

As he expertly navigated the channel for hours, Pico often spotted things that the rest of us, even JJ, had entirely over-looked. Many of the anacondas we would find in coming days, not to mention dozens of other species, were due to his sharp eye. Sometimes I'd sit in the back with him, and he'd point to the river ahead and ask, "Donde?" I would then have to guess the best route. I was almost never correct. Pico knew how to read the river with psychic accuracy: where it was shallow, where it was deep, where there would be invisible debris below that could cause a nasty collision—nothing got by him.

When Pico was only a teenager, he had been clearing land on his father's farm when a cut tree swung on a vine and landed on his back, breaking it in multiple places. He and his father made an emergency trip to Lima, where they stayed for more than a year. Several more years of recovery followed. The ordeal depleted the entirety of the family's savings and seemed to earn him a special place in his mother's heart. To this day Doña Carmen takes particular care of her forty-year-old boy.

His decreased mobility restricted Pico to certain activities thereafter. When his brothers were out in the forest hunting, climbing, or playing soccer, he was honing other skills. In this way he developed an inordinate talent for river navigation. He became the best in the area. Everyone knows that if a good motorista is needed, Pico is the man to call. He navigates the Madre de Dios for assorted causes, earning a living as he goes. His reputation is one of skill and good humor. And years of piloting the bulky *peque-peque* motors have given him a powerful upper body, compensating for his feeble legs. Pico's frame is small and wiry, but his entire torso is clad in tight cables of developed muscle. To see Pico's ripped upper body, thick neck, and clean-cut jawline as he drives, you'd never know those shriveled legs were attached to the same person.

The night after the first anaconda capture, JJ, Chito, and I left Pico to hike into the jungle to fish in oxbow lakes. In less than an hour we pulled up twenty piranha, and returned to find two pots simmering over a neat fire. Pico had made a jungle stove out of two parallel logs with a bed of molten ashes between them. A bucket was dipped into the river and lemons squeezed into its contents for a fresh batch of river-water lemonade, as we called it. The piranhas went into a pot of boiling water, onions, and garlic that Pico had prepared. Although our supplies on the expedition were few, vegetables made up most of the weight. Having items like onions, garlic, and yucca supplemented wild-caught meals and took up only a small amount of space on the boat. Items like rice, lentils, and pasta are always good ideas in the jungle, as are any foods that require only the addition of water and salt. That night we feasted. In the orange light of the fire I could see liberation in JJ's face; in the excitement of our expedition his troubles were forgotten. Here Puerto ceased to exist, as did the rest of the world; we were free.

In the morning we emerged from dew-laden tents into a frigid and overcast world. The friaje had returned with vengeance and the temperature had plummeted. Through dense fog we traveled, freezing. JJ, Chito, and I did our best to share a blanket in the narrow canoe, but nonetheless we froze. Pico was wide-eyed with concentration, at the ready each time a large beam of timber emerged from the fog to threaten our craft.

We traveled the entire second day. The fog never lifted and the cold was so miserable that there was nothing to do but make our way ever deeper into the unknown. At one o'clock the rain began. The blanket was stowed beneath a tarp, as were our packs. Our rations were light, our personal supplies even lighter; each of us carrying only a limited amount of clothing. We weathered the frigid rain mostly naked to conserve our dry clothes for the night.

For the next several days we traveled almost nonstop. Each night we retired early to our tents to escape the cold rain. Everything was damp. We caught several more anacondas similar to the first, and recorded data. None were giants, though. At night we'd have to use gasoline to start a fire on wet kindling, and would shiver and warm our naked skin while warming our soggy clothes on sticks. More than one shirt fell victim to the hungry flames. We had a good laugh over that each time.

As the cold deepened, the number of jaguar tracks notably increased. Passing dozens of beaches each day, I scanned for prints and saw indications of jaguar and tapir on almost every one. All three brothers agreed that the large cats are more active in the cold weather. Each morning we awoke to see tracks through our campsite; jaguars would come just inches from our sleeping bodies during the night. My mind strained trying to imagine it. While tending to breakfast beside a smoky fire one morning, Pico wryly theorized that it was probably the smell of fresh white gringo that was attracting them.

The frigid days were spent traveling ever onward into the wild depths. In the cold environment we saw capybara huddled on the riverbanks, always alert to danger. Spectacled and black caiman lay in places, gray and frigid in the absence of the sun, their ability to change color and shade more obvious than ever. JJ and I sat beside each other, talking at length of the jungle and life. Although unspoken, there was a feeling of celebration that our friendship had reached such heights. As we traveled he carefully imparted many things to me, pointing out species and elements of the landscape and explaining their nature.

"Looka dis," he said pointing. "You see the tree next to the capirona? That is *ohé*. My dad uses it to treat any infection." I tried to commit its image to memory, but already JJ was on to the next item. "*Mira eso*—look at this," he said excitedly. At first I saw nothing, then two eloquently plumed birds appeared from the brush by the riverside. "Gray-necked wood rail." Once again I was snapping mental photographs. Then, "Look, anhinga," he said as a bird dived into the water and vanished, or "the juice of that fruit will turn you blue," or "listen, dusky titis."

When he was not telling he was asking: "What does George Bush think about Peru?" I laughed. But on seeing his face I realized he was serious. Over the next hour I did my best to indulge JJ's curiosity about American politics before moving on to other things. "Do you have a motorcycle?" he asked. I told him I didn't but I drove my parents' car. "Is being in the military mandatory?"

As our boat was passing one beach, his head swiveled to meet Pico's eye. Moments later we pushed up against the sand and JJ hopped out, walking purposefully to what seemed to be a random point, where he stooped and began digging. "What are you doing?" I asked.

"Come see," he said, and I stood beside him as his hand delicately uncovered a cache of what looked like white Ping-Pong balls: turtle eggs. I was mystified. After days and hours of travel he had seemed to know with supernatural precision exactly where the eggs were.

Pico explained that the barely visible tracks of the mother turtle are the tip-off. They remain in the sand for a few days after the eggs have been laid, before being washed away. This, he explained, automatically ensures that a hunter of eggs would never uncover mature eggs that held baby turtles in them, for by that time the beach would have been wiped clean by wind and rain. As the trip went on he began stopping at beaches and challenging me to find the eggs. It's not as easy as it sounds.

In this way JJ and Pico collaborated as my teachers. I learned how to make rope from balsa fibers, how to gauge when the skies would dump and when they'd hold, and how to clean a kill. The guys hunted sporadically, mostly peccary, and I learned the art of barefoot tracking during the hunts. Together we'd take turns shouldering the heavy boar on the way back to the river, where we cut fronds of river cane to keep the hogs clean from the sand while they were dressed. Unlike the poachers, the Durand brothers abided by the park rules, and more important, by the code their father had taught them: only taking enough to bring home to the family. The meat was salted and placed into buckets and bags; the skin, hooves, and guts were donated to the caiman and vultures of the river.

We roasted hog testicles over an open flame and divided up the delicacy. The heads were prepared in the same manner, but these were "for the road." Through the long hours the boar's singed head would lay burnt in its pot until inevitably one of us would heft it to our mouth and use our teeth to tear off a slab of meat. Then we'd pass it. Working past tusks and fur, sucking out eyeballs, ripping jowls—it was quite a meal.

We stopped often to explore palisadas. Our skin grew taut from the endless exposure to elements. Our feet had been bare for a week. One morning I awoke before the others had risen. Thick fog gripped the land as darkness slowly lifted. The air was chilled. Across the river the canopy was a host of looming giants, dark mountains emerging from the churning mist in the half-light before dawn. Somewhere within, red howler monkeys were roaring and booming their cosmic chorus to the morning. Spared by receding currents of the La Torre, tangled lianas and towering river cane bordered the edges of the large beach. In the center of the beach in a neat line lay our row of tents, minuscule and silent among the vast green expanse. Surrounding the tents was sand christened with jaguar tracks from a visit only a short time earlier.

The morning was gray and overcast, with no sign of the sun in the sky. It was 6 A.M. *Please let this friaje lift!* I was thinking when a rustling sound to the left of the tent caught my attention. Turning, I watched as a tremendous bull tapir emerged from the river cane. Standing thirty feet from where I stood, he was a good example of how large his species can get, easily in the range of five hundred pounds. The large tapir probed the air with his stubby trunk, flaring his nostrils in all directions and surveying the beach; he took little notice of the human presence close by. Nonchalantly ambling across the sand, the large mammal left three-toed tracks as it approached the current of the river. He continued into the river and paused there to drink. He took several long gulps before raising his head and looking in either direction, and then submerged himself, save for his ears, eyes, and nose. When he reached the opposite bank he turned to sniff in our direction. Then he galloped up the steep bank and into the forest.

Pico emerged from his tent, where he had been watching my tapir encounter, and joined me. "Mucho sachavaca aquí, no?" he asked without smiling. *Sachavaca* is the local name for the species. He continued in Spanish. "This is one of the few rivers where there are so many to see. Jaguars, tapir, peccary, all animals—they disappear from the forest once too many people come. These places where no one goes," he said, now holding my gaze, "are very special."

During the long, monotonous hours on the boat my mind was consumed by the struggle to digest the incredible wealth of information and knowledge that come from life on the river. With each passing mile, the country we traveled through was redefining my perspective on wilderness, and as a result altering the entire image of wildlife conservation that had for so long existed in my mind. For the first time I was seeing a land almost entirely devoid of humanity. Back in Infierno, the community where JJ's family lived, the forests had long been hunted clean of peccary, tapir, Spix's guan, and other game. Those forests were practically empty compared to what was here. The few remaining individuals of the persecuted species were forced to lead a life of hiding. Even on the Las Piedras the wildlife was still technically recovering from a period of slaughter, but not out here on the La Torre.

One afternoon we endured several hours of violent thunderstorm that drenched everything on our canoe. Our tents by this time were bags of sand, our clothing was damp and foul, and much of the food was in similar shape. The rice had begun to rot and had to be thrown out. Our sugar was also looking poor. Shivering under the driving rain, we endured the elements for what seemed like an eternity. JJ and I huddled under one tarp, while Chito retired to the bow of the boat under another, presumably asleep for hours. So far he hadn't said more than a few sentences the entire trip; quiet guy.

We had covered a considerable distance and ascended several cataracts, steps of rushing current where the river changed elevation. The farther we went, the shallower the river grew. The La Torre is a naturally shallow channel and we were reaching the upper limits of its flow, the headwaters. Here more branches and trees lay strewn on the riverbed, waiting to catch our small launch. Often we got hung up. At that point we had endured more than three days of cold rain and endless discomfort, and when we got hung up again and again, I could feel all of us growing frustrated. One time when the boat ran aground on a submerged log, we were stopped stranded. Pico cut the motor and it was clear to the rest of us that there was only one option: into the river.

JJ stepped into the current on the left side of the boat and began rocking, but it would not budge. His eyes were dark and downcast as he submerged up to his knees and gave the boat a futile shove; it didn't move. Cursing under my breath, I too stepped into the water and joined forces with JJ to move it. The problem was that we were standing on the same log that had hung up the boat; from where we stood it was impossible to get the leverage we needed. And moving from our suspended perch on this large, submerged log meant getting fully in the deep water, which neither of us wanted to do. Both of us began to argue, and the day began to turn sour. Clearly the discomfort and cold were beginning to chip away at our patience.

Our escalating quarrel was interrupted with an obnoxiously loud assault of Spanish profanity, from Pico. Smiling as he called us little girls, he grabbed a mahogany oar and smacked it against the water, which splashed frigid waves all over us.

My eyes widened in fury. He splashed a second time, and exploded in laughter. Slapping his knee in rapture, he rocked
side to side with his eyes tearing. I lunged for him in anger, but
misstepped and plunged downward into the river and out of
sight below the water. When I surfaced Pico was riotous. Wail-
ing and hacking, he fell backward in uncontrollable laughter.
JJ dived across our pile of supplies to grab Pico's foot, or else
he would have fallen out of the boat. Chito emerged from his
tarp with a tired grin that grew into loud, hearty laughter. As
I climbed out of the water, JJ lay stretched across the supplies,
holding his brother. We were all howling now with laughter.

From that time on, the discomfort became a game. We had
covered plenty of distance and logged almost a dozen anacondas. That afternoon we measured a nine-and-a-half-footer,
and just missed catching a slightly larger specimen coiled on a
palisada. We began stopping more frequently now to explore
interesting areas of forest, or to swing on vines into the river.
We made a great deal of noise laughing when JJ back-flopped
from fifteen feet on a miscalculated launch. Later Chito was
stung on the groin by a bullet ant while peeing in the forest,
which, needless to say, also induced a fit of laughter from the
rest of us. It was a wild and perfect day.

After nearly a week of travel, when the clouds at last parted
we rejoiced. The sun set fire to the landscape, suddenly gripping the jungle in brilliant light and igniting the fantastic green
mountains of canopy around us. The change was miraculous
and occurred in only minutes. We crashed the peque-peque
into the first beach we found, and our party emptied onto its
shores to bask in the glow. To be at last in the embrace of the
sun was salvation, and we four stood with arms outstretched,
soaking it in for a long moment.

Chito and Pico began baiting hooks and launching their heavy lines into the water, while JJ and I explored the opposite beach. We found a unique area where the river had engulfed a patch of jungle the size of several football fields—from above, the river would look like a light-colored, lopsided doughnut around the dark center island of forest. Surrounding the patch of thick forest was a horseshoe-shaped beach nearly a mile long.

We walked together silently across the broad sand, observing the peculiar landscape. What made it almost eerie was the unfathomable amount of animal tracks covering the sand. It looked like there had been a parade there. The earth had recorded activity from an astounding number of species. Among the massive three-toed tapir tracks were broad padded jaguar prints. But these were only the most obvious among a myriad of others: ocelot, tayra, capybara, heron, puma, peccary, deer, tortoise, and more birds than I could guess at. It seemed as though the whole forest had emptied in recent days to stalk over the great beach. JJ's eyes were wide with fascination. Just as species do not occur homogeneously throughout the jungle, the forest itself is not one monolithic mass. As I was learning that day, there are places out in the depths that are special.

Near the water on the beach, black caiman tracks lay heavy on the beach, evidently the indication of a tremendous individual. In the last half century, black caiman, the largest predator in the Amazon, was hunted dangerously close to extinction for their highly prized leather. To acquire black caiman leather local hunters all over the basin spent decades in what became a wholesale slaughter of the species, the result being that today black caiman exist within just a fraction of their original range, most often inhabiting lakes and rivers far removed from human habitation. So as we observed the huge croc's tracks,

it was with a special appreciation and longing that we imagined what the giant must have looked like stalking over the
beach. But the find that caused even greater anguish to our
imagination was an S-shaped anaconda track almost twenty-
four inches wide. The large trunk was free of footprints or tail
drag, the signature crocodilian indicators. In my mind, over
and over were the words *it can't be*.

But there was nothing it could be. I remained skeptical, hesitant to imagine a snake so large. I estimated that a snake with
a twenty-inch stomach must have a total circumference of almost sixty inches, roughly the girth of an oil drum. *No way*.

I spent more than an hour on the beach photographing and
making notes in my field journal, recording the incredible diversity. JJ and I continued to debate the "anaconda" track.
Even though JJ was positive, at the time I was quite sure that it
was not an anaconda that had made it, although no other explanation came to mind. In the weeks to come, I would learn
how wrong I was. We were novice detectives following the
trail of a massive predator, on a path toward a discovery that
would change our lives. But on that night it was another giant reptile that would leave its mark on our expedition. The
beaches here were littered with caiman tracks that were far
larger than any I had seen, by two or three times. JJ explained
that they were black caiman prints.

Sitting in the warmth of a fire, passing around a small bottle of
whiskey, the guys told me that years earlier on a similar expedition
with their father, Elías's dog had been eaten by a black caiman.
The dog had been left on the far side of the river and tried to swim
across to meet its master. They said that the caiman batted the dog
out of the water with its tail and then smacked its jaws shut on the
mutt in midair, taking it under, never to be seen again.

That night I awoke to yelling, and bolted out of my tent to find Chito and JJ standing on the sand with flashlights in hand, whooping excitedly. Pico was on the ground laughing hysterically. The boat was rocking in the water. The massive black caiman had just stolen our fish. Packed neatly inside a pot, along with some of the salted peccary, had been the remainder of a tiger catfish. Around the large pot had been a four-foot Brazil nut bag containing much of our utensils, cooking equipment, and the scale for weighing anacondas. Pico had left the fish in the pot, in the bag, on the boat so as to discourage inspection by jaguars during the night, yet a raid had been pulled off all the same. The caiman had eaten the entire package. It took Pico days to stop bursting into laughter each time the image of the caiman's upset stomach occurred to him.

Eventually we reached a point where it was impossible to go any farther upstream. The boat was hitting sand every few minutes. Pico explained that we had to turn back or else risk becoming stranded. If the river dropped in volume, we'd have no way to return. I was reluctant to turn back; heading up that river had been the most exciting experience of my life: the anacondas, the oxbow lakes, the raw adventure of it all. "What's it like farther up?" I asked Pico. He shrugged and looked to JJ, and they both agreed that neither had any idea. "Local people only go where boats can take them," JJ explained. "I don't know anyone who has been up there," he said with a wave toward the mysterious beyond.

Progress downriver was much faster than going up—almost twice as fast, but uneventful. On what was to be the final day of the expedition, the mood on board had become once again somber. The end of our expedition was quickly approaching and we had yet to find a massive anaconda. We had caught and measured several specimens, but all of them were small. Each snake had been measured and replaced onto their debris piles, some without being weighed because the caiman had eaten our scale. Nonetheless, my notes were beginning to take on an eloquent correlation; the La Torre seemed to be proving that where humans are not, anacondas are. The only issue that continued to frustrate me was that none so far had been more than twelve feet, which is tiny for anacondas. These were males, skinny and lacking the bulk one finds in a healthy female. I spent all my energy scanning the beaches as we passed, trying to decipher every inch of the palisadas as we went, but found no anacondas. JJ too used all the power his eyes had to find even just one more giant snake, but found nothing.

At one point I was beginning to doze when the canoe suddenly shifted as Pico sat bolt upright. Before anyone said a word we all looked to the mad motorista; we knew he had locked on something serious. His eyes were wild with frozen awe as he pointed to a large palisada in the distance. Our heads spun around to the pile of forest debris downriver from us, where a tremendous pile of anaconda coils lay basking. It was by far the largest snake I had ever seen.

Action enveloped our tiny vessel. Because she was in the center of the large palisada, we decided to split up. I would come from upstream while JJ and Chito came from the other side. Pico maneuvered the boat to about seventy feet upstream from the snake. In the turbulent current I had only a split second to jump onto land. "Cuidado tus huevos—*Watch your balls!*" Pico grunted through a cigarette grin.

The palisada the anaconda had chosen was the biggest we had seen, stretching for nearly the length of a football field and rising high above the river. Looking downstream, I peripherally noted Pico and the others moving into position, but my eyes remained glued to the snake. She was spellbinding even from a distance. JJ and Chito were climbing out of the canoe, and I took a step toward the snake, which was still seventy feet away at least. Ever so slightly the muscles of her powerful coils tightened and her head emerged. She was awake.

Without any warning she lunged across her body in the direction of a large, hollow tree. I broke into a wild sprint down the trunk of the fallen kapok, screaming for the others to let them know that the moment had arrived ahead of schedule. Her body was disappearing so rapidly I was sure she was going to escape as I rocketed over the timber mountain toward her.

Barefoot and naked save for briefs, I closed in. When I reached her, only a third of her body was left visible; the rest had disappeared into the timber below. First I tried lifting her heavy tail, but I quickly realized there was no chance of pulling her back. My only option was to climb down with her.

The anaconda had slithered down the shoot of the large hollow log that sloped through twelve feet of debris and into the water. Scrambling and sliding down the top of the log, I slid into the bowels of the wooden mountain. Landing with a crash in muck and thorns, I found myself on the floor of a cavern of packed wood. The way the timber had piled up in this spot had left a large gap, almost like a hidden room, and I was now inside of the mountain, face-to-face with a fifteen-foot snake.

Looking up through the tunnel of the great hollow log, the anaconda dominated the scene, her coils pulsing and alive with ripped muscles. She was steaming fast toward a gap in the timber where the water showed below. With heart pounding I screamed one last time in vain for JJ and Chito, and went for her head. I had barely taken one step before she spun around in a sweeping arc of a strike, mouth exploding open 180 degrees to reveal a sixteen-inch gape. Mottled purple and black and lined with six rows of needle-sharp teeth, the tremendous jaw swept across the cavern toward my face.

I collapsed backward to evade the strike, falling flat to the ground. The snake's head sped past the open air where my face had been and continued on, carried by the momentum of the lunge. Thankfully, anacondas of such large size, while agile in the water, are plodding on land, slowed by their own bulk. The momentum of her strike launched her front half violently across the cave, and when she landed I knew that there would not be another chance. I regained my feet and pounced, securing a grip behind her head. Big mistake.

Instantly her body began to twist and curl, binding my arms together between powerful coils the girth of my thigh. Crushing with indescribable power, her trunk wrapped once, then twice around my arms, threatening to snap the joints of my elbows. Her weight forced me to my knees and I screamed in pain. The moment had gone from hectic to out of control with amazing speed, and I was alone. As another of her coils rolled up over my shoulder and around my neck, it became clear that even if I released her head, there was a very real chance that this snake would crush me to death.

The level of force from a constrictor of her size can in seconds flatten the rib cage of a caiman or tapir, animals much tougher than humans. Panic surged in my veins as her impossible strength mashed my shoulders toward one another, flexing my clavicle like a twig.

It was then that JJ came into view, hurtling at full speed

down the same log I had used. The momentum carried him to the opposite side of the cavern. He swung around with knees bent, arms spread, and mouth agape in shock. "Whoa!" he shot out, paralyzed for a moment by what he saw.

He tackled the trunk of her coils and protected my neck as I tried to maintain my grip on her head and the snake beat us against the jagged walls and floor, writhing in alternating directions. As more and more of the anaconda's body emerged from the hollow log, JJ and I continued to wrestle the front end. With the final third of her body came Chito, who had been dragged *through* the hollow log, grasping her tail. He was covered in yellow and white snake shit, buckets of it.

Screaming and straining, we began heaving the coils off my arms. It took us fifteen minutes to get control of the snake, and even longer to get the snake up out of the cavern in the great pile of wood and to the edge of the island. Pico maneuvered the canoe into position below the palisada and the four of us (anaconda included) dropped into the tiny canoe. Pico's eyes were bulging out of his head as the canoe nearly capsized in our struggle to control the giant.

On the opposite beach we got out and collapsed. The snake was unrestrained but both humans and reptile were too spent to move. The three brothers kept their distance from the anaconda as I knelt in awe. She lay curled in a tight spiral, peacefully protecting her head, as I soaked in the sight of a creature I had always dreamed of seeing. Legendary in both size and spirit, she was the find of a lifetime. She was magnificent, large green and black scales on her back fading into vivid yellow and black toward the stomach. Emerging from her mouth at intervals was a thick, pitch-black tongue the width of a man's finger, which probed the air with regal indifference.

We took measurements and a few photos, careful to minimize the time she'd be handled. She had an average circumference of twenty-eight inches, and a total length of fourteen feet, nine inches; what a beast! Her even shape, well-toned muscle, healthy color, and glossy scales indicated that this snake was in top physical condition. During our inspection the anaconda was placid and cooperative, not at all defensive, let alone aggressive.

With measurements and photos taken, it was time to release her and end the stressful ordeal. JJ, Chito, and I carried her on our shoulders, gently placing her near the water's edge. Incredibly, she didn't bolt. For a time the snake lay on the beach and I sat beside her. My hand rested on the cool scales of her thick trunk, which rose and fell as she breathed. I tried to carve the sensation of her, the image of that massive body, into my mind. JJ and crew hung back, still wary of the giant and nervous that her unrestrained jaws were so close to my face. At last she began to move, sliding en masse into the water. My hand remained on her back as the length of her tremendous body slipped beneath the river's surface with supreme grace. Then she was gone.

The final hours of our expedition were spent splashing and frolicking in the river, then cooking, laughing, and recounting stories around a fire, long into a night of a million stars.

10

The Floating Forest

A much more important problem is the question as to the existence of the carnivorous monster that has left its traces in the glade.

—Sir Arthur Conan Doyle, *The Lost World*

The magic of that hulking female anaconda stayed with our expedition for the two days it took us to return down the Tambopata. Our emotions were soaring in the wake of such overwhelming success. I still couldn't believe all that I had seen and kept switching on my camera and checking the photos to reassure myself that it had all been real. Pico's river-navigating lessons gave way to practice, and I piloted the boat as we arched with the current. JJ's face bore none of the twisted frustration of the week before. Even Chito had loosened up after our adventures.

We arrived at Don Santiago's farm at dusk. He was sitting beneath a palm shelter beside the river, drinking soup. His sons greeted him with hugs and I received a friendly nod. He motioned us to sit and I noticed a stark change in behavior among the brothers. Like privates suddenly in the company of a general, they spoke and even moved differently around their father. And so did I.

As we stampeded in, he rose to put a log on the fire, and for the first time I was able to inspect the octogenarian. He was dressed in a filthy buttoned shirt and trousers. Most days he wore rubber boots but today he was barefoot. Even so he stood at about the same height as me, roughly five foot ten, but bent and sinewy, with misty eyes and large, powerful hands. Placing the log onto the red coals, he blew with expert precision; several flames leapt alive in the cracks of the wood. He swung an ancient coffeepot onto the wood and then sat again, pushing out his chin and dragging his sausage fingers across coarse porcupine-white stubble. Watching him move, I couldn't help but think of my grandfather, my dad's dad, who was the same age but a very different creature. Living alone in the jungle at just a few years shy of ninety would be a death sentence for him and just about anyone I could think of back home, but not Santiago.

Santiago had grown up in old Puerto Maldonado, back before it was really anything, when there were only about a thousand people living there. In those days it was a tiny settlement, just an outpost on the river, really. There he lived on the banks of the Madre de Dios River in what is known as the Old Town, Pueblo Viejo.

His mother died when he was fourteen, and so parentless, penniless Santiago took to the river working as a transporter, piloting canoes up and down the Tambopata River. This was long before Puerto Maldonado was connected by road to the outside world, in the days when the only way to get in or out was a rare floatplane journey, or river trip. As a result, all goods coming in or out of Puerto had to come and go by boat; and in those days, there were no motors.

Young Santiago used his athleticism to pilot canoes by pole, against the infinite resistance of the Tambopata, almost a hundred grueling miles upriver, all the way to the state of Puno, where donkeys loaded with supplies from the outside world would meet them to exchange goods. Santiago came of age in the wild west of the Amazon at a time when that corner of the earth had been abandoned for centuries, a sleeping giant. Though he had not been a native of the lowlands by birth, his time on Tambopata launched him deep into the world of several local tribes that ranged from peaceful known communities to fierce uncontacted tribes.

Today the upper Tambopata is among the most pristine forested wild places on the planet; back in the days Santiago was poling canoes, it was virtually untouched. As an adopted native, Santiago learned the language of the Ese-Eja Indians, as well as their medicines, beliefs, and insights into the vast jungle. He encountered jaguars and giant anacondas, and some species that have yet to be described to science. Before he was twenty he had undergone an initiation ceremony in which a shaman sewed the nerve of an electric eel into his forearm, guaranteeing strength and virility for life. From then on, he was native.

As he grew into his twenties Santiago joined the region's police force. It was a time when the gruesome process of religious conversion was in full swing and missionaries were still constantly striking out toward remote places to convert the locals. Santiago followed these expeditions as a police officer, sometimes traveling for weeks on end to reach remote locations. He was always in a protective position, ensuring that the inevitable conversion happened in the most humane way possible.

An accident with a falling tree in his mid-twenties crushed Santiago so badly that it took him almost three years to recover (an event that was eerily mirrored by Pico later on). In order to get the medical care he needed, he was sent out of the jungle and over the Andes to the coast, to Lima. There he connected with his grandfather for the first time. Santiago's grandfather made use of the time to send him for more schooling, and so Santiago, a backwoods boy, was educated in sociology and anthropology.

After he returned to the Madre de Dios he worked as a policeman for several more years, always in the jungle, and always with indigenous people applying his compassion and newly acquired education. When he quit the police he was still a young man, now with a good pension. He dived into working to better the life of the people who had taken him in when he was young, the indigenous people of the Tambopata. He settled on the river and began a farm. And with seven others, he began organizing the clans and communities that lived along the banks of the great river. After years of work and applications to the government, Santiago and several others founded the indigenous community of Infierno in 1974. This gave the people of the lower Tambopata something they had never had before: their own titled land, recognized by the government.

In the years that followed Santiago began a family and, according to JJ, swapped five hundred hectares of land for his wife. As they began having child after child together, Santiago continued to improve the community of Infierno, even setting up a school and a medical post there. Thus JJ and his brothers were born in the jungle. When they were old enough they would travel to Puerto Maldonado for school, but always, the jungle was home.

JJ tells the story of the day two boys floated down the Tambopata on a raft. They were from the Andes and had escaped a terrorist organization similar to the Shining Path. When they arrived at Infierno one of the boys took off; the other was taken under the wing of Santiago and Carmen. Years later, when the young boy, who would take on the name Alex, noticed in Puerto members of the terrorist group he had escaped, Santiago took the matter to the police. The tip from Alex resulted in a huge bust, and the government of Peru congratulated Santiago and his new son in an official adoption ceremony. Later on, I would know Alex as the most skilled birder and guide I have ever met.

And so Pico, JJ, and Chito, three jungle boys, proudly recounted our La Torre expedition to their father, who listened politely. Every so often he would nod and say, "Claro, claro," for "I understand." When we got to telling him about the biggest of the anacondas, Santiago laughed hardest at the parts when Chito got sprayed in shit and the gringo almost crushed. "It was probably half the size you say it was," he said with a dismissive wave of the hand.

JJ excitedly said, "Tiene fotos, papa! Mira aquí," and reached for my camera.

I pulled up the photos of the biggest anaconda we had wrestled, the large female. The old Amazonian regarded the photographs. Then he began to shake his head. "You know," he said through a frown, "this is the smallest anaconda I have ever seen." Pico cursed and slapped his hand on the table, laughing. JJ spat. We were all shocked. He placed the camera on the table. "I have seen anaconda eating tapir before, anacondas this big!" He held his arms in a hoop to show us the dimensions of the snake from his story.

Now it was my turn to laugh. What he was describing was impossibly large. Try holding your own arms out, joined at the fingertips, and make the biggest circle you can. The snake Santiago was describing was bigger. *No way could there be anacondas that size*, I thought. " If you are crazy, and clearly all of you are," Santiago hissed in a rasped whisper, " if you want to go see big snakes—real giants—then go to the *aguajal*."

JJ and the others fell silent and I made the error of letting the incredulous smile from moments earlier hang too long on my face. Santiago turned his attention on me. "You laugh?" he said, suddenly lifting the ancient folds of skin from his eyes with his brow and looking directly into mine. "Let's see you laugh when a forty-foot snake is swallowing your ass."

The candles at the center of the table illuminated our conversation long into the night. Santiago told tales of expeditions past, anacondas, jaguars, tribes, and other lore. We laughed and drank long into the night, until Santiago finally decided to turn in and stumbled drunkenly off into the jungle. When he was just out of sight, we heard a sharp streak of swearing as his bare foot landed in a pile of pig shit. The three brothers fell to the ground, overcome with laughter, as their father shouted threats through the bush; you could hear the smile on his leathery face. Before sleeping myself, I savored the misty stillness of the night and spent well over an hour attempting to commit everything I had heard to paper.

JJ and I were determined to get to the anaconda lake as soon as possible, but it took us a few weeks. In the days that followed the La Torre, JJ and Pico and I ran a volunteer group up to Las Piedras. But the glow that had swaddled us after the La Torre expedition dissipated after a vicious fight between Emma and JJ. The problems were only deepening and money was disappearing faster than ever. The difference was that by then it was no longer only their battle. I was in it now. Bringing groups, recruiting, and spending every ounce of my consciousness and resources to help save Las Piedras Station had become the norm for me. But it was not enough.

On the night before we brought the volunteers upriver, Emma, JJ, and I had met at their house to recheck the expedition supplies and go over final details. The tension between them was so palpable that I wanted to leave the room. Over the course of an hour the conversation stayed artificially civil, but then it finally broke. In escalating Spanglish they moved from room to room in passionate debate. There was no way for me to help, no way for me to leave, and no sounds in the still night except for their voices. It went on until Joseph, then three years old, woke up crying. The interruption left a silence that was worse than the din of moments before. JJ's eyes were slits of dark passion; Emma's were dry and wide, as if in defiance of the circumstances. She looked at the top of JJ's head for several minutes as he looked at the floor, before she said softly, "We have to sell it."

Since that night nothing had been the same. JJ and I ran the expedition on autopilot. After two weeks we returned to Puerto, where JJ and I stayed in a hotel room for a night, since things were ever worsening between him and Emma.

On the day we left to find the mysterious lake that Santiago had told us would hold the anacondas of our dreams, I was far from mentally present. As JJ and I moved from car taxi to boat to trail, we were barely distracted from the fact that the station might be gone soon. I could not wrap my head around it, or shake the despondent gloom it had thrown over me. There had to be some way to help them more, some way to bail them out—get more business to the station. I just didn't know what it was. I wrote letters to prominent conservationists, sought funding, but at that age I essentially lacked the skills to do anything effective to help the situation.

First it had been the Brazil nut farmers that wanted to steal the land, then another, more powerful group had tried to steal the station itself. Most recently, an oil company had barged into the scene—even landing their helicopter on the beach adjacent to the station and setting off explosions within the reserve. Fighting on so many fronts, and against foes as powerful as an oil company, had run Emma and JJ down fast, and it was only getting worse. And although I was bringing groups to earn money, and even chipping in out of pocket here and there with savings from lifeguarding back home, my contribution to the effort was still hardly making a difference.

JJ was also not himself. At one point while traveling he sent me ahead with his brother José, promising he would follow in a few minutes. I waited at José's house all day: over nine hours and no JJ. Angry and frustrated, I finally put up my tent that night, resigned to the fact that JJ wasn't showing up. But as soon as I had crawled into my sleeping bag he appeared out of the jungle. "Let's go!" he said energetically. I had seen him do this before: follow up an irresponsible act with a gush of enthusiasm and smiles to cover his ass.

"I'm going to sleep, JJ," I groaned from my tent. I was furious at him for leaving me all day. "Where were you?" I asked, and he giggled and told me I should have some coffee. I had not yet learned that sometimes JJ disappeared for long periods with no explanation. It was maddening, but some minutes later he convinced me that he was sorry and that we should still go explore the lake, despite it being night. We began walking the long, straight path toward the lake of Santiago's legend.

During the course of the hike we encountered night monkeys, many frogs, and a tapir. With each encounter my frustration slowly dissolved, pushed to obscurity by a growing sense of anticipation. I reasoned that there was no chance of finding an anaconda tonight. I had too much experience searching the jungle to be so foolish; but that little voice in my head, or maybe my gut, had begun to glow warmly. I had a good feeling about this.

The trail was straight and our march long and uninterrupted through an endless tunnel of trees. José was with us. He carried an old rifle over his shoulder, hoping to pick off a paca or other game for the pot. After more than an hour of walking we reached where the main trail turned and dropped into the floodplain. There we stopped for a moment, each of us choosing a different place to pee. As I was unbuckling my belt, I noticed a blue glow at the limits of my headlamp's beam. *What the hell is that?* I wondered.

Crawling through the dark foliage on my hands and knees, I went after the hypnotizing light that shone from the detritus. Pushing branches aside, I felt my heart quicken. What could be shining like that in the forest? When I reached the spot I fell to my knees. Before me were the glowing wings of a blue morpho butterfly. There was no body, just the four wings. The metallic cerulean iridescence was throwing the light of my headlamp back at me; that's what I had seen. I lifted the wings with tremendous care and made my way back to the trail. José saw me staring into my palms and approached. He whistled in surprise.

Blue morphos are large, fast butterflies, and it is rare to get a close look at their stunning wings. There are myths that the morpho is a spirit or enchantress that tempts men to follow and then brings doom. Others believe that it is a powerful spirit that, if caught, can cure the ailments of the capturer. Some believe they are messengers of the forest gods. Standing beside me, José gazed at the wings for some time and then cupped his hand under mine. "This is very lucky," he said, looking into my eyes.

He turned and searched for a moment, and then cut a large

leaf. Taking the wings from my palm, he laid them on the leaf with reverent care, and then folded the leaf into a package, tying it off with root fiber. "Keep this," he said, handing me the green parcel. For some reason I thought of my uncle Albert, an old Catholic priest, giving me rosary beads as a child.

We descended the floodplain slowly and in silence. By now there was a kinetic energy to the night. I am not a superstitious person, but finding the wings had cast a spell, a feeling, of luck onto the night.

At the foot of the terra firma began the gnarly floodplain, overgrown with thousands of aguaje palms and pocked by holes of varying depths. "Mucho cuidado. Be very careful!" José whispered as JJ fell up to his waist. As we walked it was impossible to judge the depth of the mud. Some places held our weight; in others we were over our head. Progress was slow. "Is this where the anacondas are?" I asked, but José said no. We were close, though. The aguaje palms ended and suddenly we were in a great clearing.

All of us turned out our lights, and the brilliant moonlight revealed a surrealist landscape that bound us all in transfixed confusion. It seemed to be a field. No, there was water in places. The trees were all disorganized: the tops of some palms spread directly out of the ground, and others grew tall; there were smaller trees growing from the field, a kind of dwarf forest. I was struggling to understand the sight while José stepped into the brush. He cut a tall sapling, maybe thirty feet, and trimmed the branches from its top. Then he walked to the edge of a water area and thrust it downward. Gravity pulled the long pole through his hands into the water: ten feet, fifteen, twenty, twenty-five, and then it was gone. JJ whistled in surprise. José turned to us. "It's very deep and very dangerous. We are here."

JJ and I exchanged looks of wonder. The field was not a field at all. It was floating grass. The reason some palm trees looked like their tops were sprouting from the "field" was that they had grown from the bottom of the lake, and we were seeing their canopy. Suddenly it came into focus.

Cautiously exploring, I walked to the edge where José had demonstrated the depth, and saw that a fallen palm tree stretched across the water, half floating on the water between the land and the grass island. "Can I walk on it?" I asked. Alarmed, Jose answered with an emphatic no. But I was unconvinced. Leaving my camera and other valuables on dry ground, I was able to balance on the fallen palm and walk out across the water toward the grass. José sputtered and cursed in surprise. "Come back!" he whispered. (We were all whispering, for some reason.) But I made it to the grass and stepped off the log. The ground sank under my weight, the way a raft does. One foot, then the other, and I was standing on an undulating, soggy mat, floating on the surface of the lake. "Look," I whispered, "it's fine!"

Just at that moment the grass gave out, and I plunged straight down. Suddenly I was underwater. From the terrifying, cold blackness I reached up and grasped at the grass above. Desperately I pulled my self back up toward the air. When I resurfaced, JJ's hand grasped my arms. I scrambled and he pulled—both of us were working to get my body out of the predatory black abyss below. "Dios mio!" he said, smiling. "I thought you were gone!"

Now we were both on the floating mat of grass, above the lake, amid the canopy of palms. Cautiously placing our first steps, we discovered that staying by the dwarf trees made for better footing, since each tree had a clump of roots and grass at the bottom that was sturdier than the areas of just grass. The moon was so bright that we could explore without the use of our headlamps.

"Are you coming?" JJ asked over his shoulder to José, who essentially replied that there was not a chance in hell. "See you in one hour," JJ hissed back into the darkness where José stood.

With that we began exploring in earnest. The terrain buckled and gushed under our feet, bobbing and bubbling, sinking, and tangling our limbs as we slowly moved across the archipelago of grassy islands. We passed the tops of some trees, and the bottoms of others. In the eerie moonlight spiderwebs were ghosts all around us. More than a dozen times I caught dark patches in the water and turned on my headlamp to reveal the red gleam of caiman eyes watching us from the motionless obsidian surface of the lake. An owl surrounded us in its ominous warning. Omnipresent and unknown eyes of hundreds of creatures, obscured by darkness, watched our progress.

This was farther down the Amazon rabbit hole than even JJ had ever been. It looked like we had reentered the Jurassic. The blazing moonlight only accentuated the phantasmal archipelago of floating islands. This was Dr. Seuss on acid. "I feel like I am on ayahuasca," JJ said, looking around after forty silent minutes on the strange terrain. "Me, too," I whispered, though I had never taken the hallucinogenic at that time.

"Look-a-dis," JJ said, pointing to the ground. I turned and stepped over to where he was looking; he traced his finger along an area where the grass was pushed down in a large S-shaped trail as wide as my waist. "What the hell is that?"

I asked, though I already knew what he was thinking. "This is anaconda," he said slowly. "It's a track." "No, it's not," I replied quickly; it was far too big.

Yet as we walked in the hours that followed, completely absorbed in exploration and forgetful of José, we saw many such pathways, all in sweeping S shapes. Some were as thick as my arm; others had the girth of an oil drum. There was no way, I thought, that these could have been from snakes. *No way*, there were too many of them, they were too big, and they seemed so fresh . . . It was as though as we walked, dozens of giant snakes were slipping into the water, just out of view.

Without warning JJ vanished. Suddenly alone, I stared bewildered for a moment at the hole in the grass beside me. Then, as my brain caught up, I dived onto my stomach and reached down into the dark water. From below the surface my hand found JJ's. I pulled and JJ scrambled now, and within moments he emerged soaked and panting. We were both shaken, and in that moment realized how utterly terrified we were of our surroundings. It was several moments before we began to laugh hysterically.

From that point on we stuck close together, helping each other along. We were discussing caiman behavior and picking our way along a large section of especially perilous grass when JJ froze. Suddenly, grabbing my shoulder with one hand, he pointed toward the ground.

What looked like a beached whale was lying not five feet from us. I could actually feel the blood drain from my temples as synapses struggled to register what my eyes were seeing. All the while we stood motionless, bobbing gently up and down for a long moment of slack-jawed silence.

The snake before us was lying flat on the ground, but her

back was as tall as our knees. Her gargantuan shape spread out in sweeping curves into the darkness in either direction, her fat midsection just feet from where we stood. Vision came in waves of comprehension instead of all at once. This snake was easily double the size of the one we had caught just two weeks earlier. This was a monster.

JJ's hand was viselike on my shoulder, while with the other he pointed to where my eyes had also just traveled: the body of a *second* anaconda. Without taking our eyes away from the tremendous snakes, we stood there and hugged in triumphant awe. A smile was forcing its way onto my face. This snake was as thick as a small cow, and easily well over twenty-five feet long. This was not just a large snake, but the mega-snake of legends. Somewhere in my consciousness, the thought fired that if we could get a photo of this snake, we would be on the front cover of *National Geographic*. In all the years of daydreams, all the months of trudging through the swamps at night, I had never imagined that reality could offer something like this.

We were silent, both of us basking in the fantastic wonder of the moment. Each of her giant scales glistened, smooth and in perfect symmetry with the thousands of others that covered every inch of her tremendous length. Her immense head lay partially obscured by the tall grass, her back broad and smooth like that of a whale. I could see her watching us, sampling the air with a great black tongue, itself the size of any ordinary snake. The second anaconda was considerably smaller, perhaps fifteen feet in length, and lay draped over her in several places, seemingly in amorous embrace; both lay still, as surprised by our presence as we were by theirs, and waiting to see what would happen next.

When we had stopped walking I had frozen in midstride, stunned by what I had seen. Now, however, I placed my foot

The snow-capped Andes abutting lowland tropical
rainforest. *(Renata Leite Pittman)*

First views of the Las Piedras River.

All photographs courtesy of the author unless otherwise noted.

An indigenous community on the banks of Las Piedras.

The main deck of the Las Piedras Biodiversity Station.

My first moments with Lulu.

Field notes: walking through the
forest with my anteater.

Hammock view of an anteater's afternoon nap.

JJ pulling the boat past a waterfall,
giving careful instructions all the while.

Pico, the mad motorista, with his
cane, rifle, and signature smile.

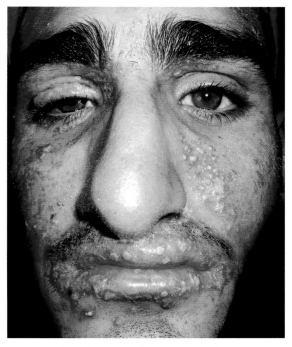

The infection: After seeing my face,
I was sure I was dying.

Peccary heads, macaws, spider monkeys, and many other forest
wildlife were found dead in the hands of poachers.

The seemingly endless La Torre winding through
the surrounding Bahuaja-Sonene National Park.

JJ and me with a heavy female anaconda.

Despite their fearsome reputation, anacondas are quite docile
when they do not feel threatened. In this video still,
a healthy fifteen-foot female allows me to inspect her.

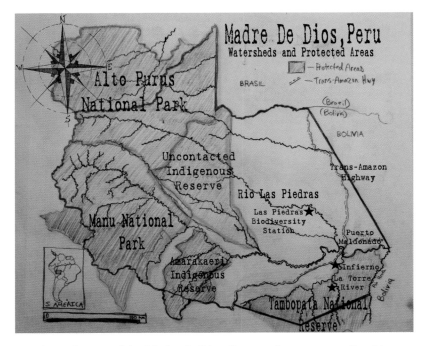

Field notes: a diagram of the floating forest.

A rough map of the Madre de Dios. Protected areas are outlined in green, and stars mark places of note. *From top to bottom*: Las Piedras Station, Infierno, La Torre.

Field notes: the mother of them all. On our first night out in the floating forest, JJ and I encounter a massive female anaconda.

Burnt wreckage in the wake of the Trans-Amazon Highway's
creep toward Las Piedras.

The highway grants loggers easy access to the
untouched forests of old-growth timber.

The skull of a howler monkey adorned with macaw feathers, in Santiago's hut. It was there that I first learned of the floating forest, the Western Gate, and the legendary anaconda with horns.

While I was lost in the maze-like swamp, the jaguar came within inches of me in the night—close enough that her breath warmed my ear.

A fourteen-foot female anaconda constricting a peccary.
(Gowri Varanashi-Rosolie)

The sweeping bulk of the Tambopata River. *(Tom White)*

An angry female anaconda twists and turns, throwing Gowri, JJ, and me around as we attempt to measure her. You can see her sinking a few teeth into my elbow.

Gowri's first time at Las Piedras.

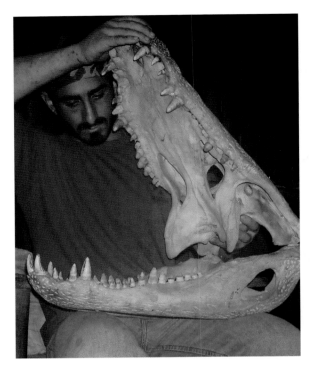

The skull of a black caiman.

Macaws, parrots, and other animals regularly visit the colpa located in Las Piedras.

Video still: a small tributary several days' journey beyond the Western Gate.

Beyond the Western Gate, alone, in the deepest jungle on earth.

Field notes: seven people versus one anaconda in the floating forest.
The giant took us down.

delicately down, and the vegetation rocked gently in the water. A tightening of the snake's muscles became discernible. Her head rose from the grass off to the left, flicking a tongue as thick as a man's finger up and down, scenting the air. Then, all at once, she bolted.

My brain fired a hundred thoughts all at once as her coils exploded into action, rapidly entering the water and disappearing. Propelled by an irrational urge to restrain the snake and get photographic evidence of her size, I dived onto her back like a shortstop catching a line drive. My presence did nothing to impede her progress, and my arms could not close around her, such was her circumference. I was carried more than seven feet *on* the anaconda, my arms clinging to her trunk, legs dragging along beside.

As she swept my body over the surface of the lake I tried digging my heels into the grass but they never held. For the first time it suddenly occurred to me that if she opted for fight over flight, then that giant tooth-filled mouth would do irreparable damage to my face, not to mention that she'd be able to collapse my ribcage in about half a second. Thankfully, she never struck. Instead her head entered the water, followed by the first third of her body, and then the section of her body I was gripping. Digging both knees into the grass, I tried to brace. But her power was as unyielding as a horse, or a truck: there was *nothing* I could do. She dragged me fully into the water, face-first, as JJ watched in frozen astonishment, his circuits too blown to move.

Cool black water swallowed my head and shoulders in the flash of an instant as the rest of my body, clinging to the giant snake, followed into the dark abyss. I let go and paddled like a windmill for the surface.

With my head above the water I held on to the nearest grass,

where the last fifteen feet of the leviathan trunk sped over my
shoulder and through my hands into the water. In profound
awe I carefully savored the last moments of her presence, hands
recording with every fiber of their ability the smooth, scaly
immensity passing by. With my heart jackhammering in my
chest, it was a moment of unparalleled shock; I was intimately
at the mercy of perhaps the largest snake living on earth. As
her body narrowed to her tail, my hands came closer together
and then finally touched as she departed. My last view of her
was of her tail disappearing into the black water below.

That first night in the floating forest became the stuff of
legend. After the giant female had left, we caught the smaller
snake, a fifteen-footer, and brought it back to the edge to show a
terrified José. Though we told him what had happened with the
giant, he did not believe us. No one did for years to come, and
that was fine; what we had stumbled onto that night was a place
of secrets unparalleled in the Madre de Dios or the rest of the
Amazon. In years to come I would query scientists, local people,
experts, anyone I could find, but none of them documented anything like such a place or seen such a snake. For JJ and me,
it was the beginning of a whole new adventure, a relationship
with the floating forest that would yield many more encounters.

It was sunrise by the time we got back to camp at José's,
where we fell into our tents for much-needed sleep. That night
we sheepishly returned to Don Santiago's hut. There we told our
tale to an eager Pico and a smirking Santiago. JJ reenacted my
jumping onto her back as he stood there with his mind blown. "I
told you," said Santiago, smiling tooth and gum. Like the other
times we had stopped at his hut, on this night we sat for many
hours talking, eating, drinking, smoking, and telling stories.

Though Santiago had not been as fast as the rest of his family to accept me, in the wake of the La Torre and then floating-forest expeditions, it seemed at last I was in. Santiago dug his teeth into the skull of a peccary head and then passed it to me. Like Pico and JJ, he seemed to find this forest-loving gringo a curious thing; it just took him longer to open up. Long after JJ had retreated to his sleeping bag, Santiago and I sat together talking for the first time alone, and then we sat together in silence.

I was aware of him watching me. Throughout his many decades, Santiago had seen creatures yet undescribed to science, tribes that no one knew existed. I knew I was in the presence of a living encyclopedia of the Madre de Dios, a man with a lifetime of hard-won secrets, though I do not know what prompted him to tell me what he did.

Over the soft crackle of the nearby coals he said, "There is a place I saw once that no one has ever been to." I looked up. He continued in his gravelly whisper, "It is a place wilder than any other I have seen."

"Like La Torre?" I asked. He shook his head and spat out some cigarette paper.

"La Torre is nothing compared to this," he said. His hand was flat over the table now, indicating a pause so that he could be certain I understood what he meant. "La Torre is a place that very few people go, it's very wild, but this is different. What you don't understand is that even one or two people change the spirit of a place—the animals change behavior. The place I am talking about . . . no one has been there for centuries."

"What's it called?" I asked dumbly as he poured newly heated tea into the mug before me.

"It doesn't have a name," he said without smiling.

I tried not to let my eyes fall out of my face onto the table as he spoke. He told of watching a mother jaguar play with her cubs on an open beach, and of otters teaming up on a twenty-foot black caiman. There harpy eagles haunted the canopy and flocks of macaws filled the sky like flying rainbows; the river was so thick with fish that you could scoop up dinner with your bare hands. The only tribes who knew of the land had regarded it as sacred and never entered, and so the place had remained untouched for millennia. What he described was a lost world.

He explained that it was not reachable by boat, that a person could only get there by crossing from one river over to another, and then hiking upstream. "There is a huge tree, like a *puerta*."

"It's a door?" I asked, confused. I interrupted him only when my Spanish failed me, or at times when he used Ese-Eja instead of Spanish by accident. It took me a while to understand what he meant here was that the tree was a barrier, a gate that prohibited any chance of boat travel, even by canoe. What lay westward was unexplored. "La Puerta Occidental," or the Western Gate. He offered no further explanation.

When his story finished we sat in silence for a long while, as my mind struggled to comprehend what I was hearing. The La Torre had shattered everything I'd known with its stunning abundance of wildlife. How could there be a place even more pristine? We must have sat in silence for thirty minutes before I worked up the nerve, slowly. "Show me."

At this Santiago chuckled. "When you're older," he said, as he stood up. "Good night," he announced, then left the thatched hut.

"Good night," I said in return. After that I was alone, left with the hopeless task of sleeping as the sensory aftershocks of riding an anaconda still echoed on my skin and the wonder of Santiago's stories coursed through my brain. For hours in the darkness I watched the Tambopata slip beneath the moonlight, and wondered what magical world I had entered.

BOOK TWO
THE BATTLE OF
THE AMAZON

11

The Other Side

Fate gives all of us three teachers, three friends, three enemies, and three great loves in our lives. But these twelve are always disguised, and we can never know which one is which until we've loved them, left them, or fought them.
—GREGORY DAVID ROBERTS, SHANTARAM

By the time I reached the age of twenty-one the Amazon had become the central, driving force in my life. The worries of my teenage years had long faded away: I had not been born in the wrong century and missed out on the age of adventure; instead I had just been ignorant of where to look. In the span of three years I had caught crocs and anacondas, raised an anteater, begun working as a conservationist protecting the place I loved, and become friends with JJ and his wild, indigenous family. Then, of course, there was the discovery at the floating forest, an event that radically restructured my view on the world: if a place like the floating forest and a creature like the snake I had ridden could exist, what the hell else was out there?

When I was home in New Jersey, my entire identity was

the Amazon, and I liked that. There was never a time when I didn't wear my peccary-tooth necklace on a balsa-bark cord. It was a great recruiting tool: I'd walk into a deli or outdoors store or go on a hike and inevitably someone would ask what animal it had come from. Many conversations would start in this way, and as a result, many people came to Las Piedras.

I was getting talented at leading a double life. During college semesters I would work and save for the next expedition, as well as plan, recruit volunteers, and explore from an academic angle the things I had experienced in the field. I'd plan with professors to take my finals early, and a week before semester's end I would pack my bags and head to JFK Airport. It was like jumping between worlds. You drive to the airport on the cold concrete, past leafless trees beneath a gray sky. You journey for dark hours high above the earth, before being spat out the other end of the wormhole into a boiling tropical wonderland. In one life I had a cell phone, checked email, studied for tests, and watched TV at night; in the other I drank from a river, hunted for anacondas, and explored the secrets of the earth, sometimes in places no one had ever been.

Returning from the Amazon stained with the red earth of the jungle and stoked with the flame of high adventure, I found my life elevated to a level beyond my wildest dreams. Yet it was not without struggle. Of course, there was the constant cloud cast by the trouble with the station, and often I would send part of my meager lifeguarding paycheck to Emma and JJ, to help with the cost of legal documents, travel, and the other minutiae of defending what was ours, a fact I discreetly withheld from my parents. I also wrote letters, made phone calls, and sought the help of anyone who might be able to help protect a research station in the Amazon—but I made little to no progress.

My mom and dad were my greatest supporters. They could see that in the jungle I had found the crossroads of passion and purpose, and they seemed pleased about it. However, others questioned my path. Some extended family members and friends would ask me when I was going to "grow up" or "find a real job." Somehow, to them what I was doing seemed childish. One relative repeatedly asked, in an exasperated tone, just how many *vacations* I was going to take to the jungle. On another occasion a friend of my parents described his new mahogany bed set with enthusiasm as I squirmed.

I also found myself struggling for acceptance in college. Although I had great enthusiasm for ecology, conservation, and the many strata of disciplines related to protecting and understanding Amazonia, my dyslexia and poor academic performance earned me the scorn of more than a few professors. In return, I developed a noted distaste for the dry and joyless way in which many of these accomplished academics plodded on through their work. Teddy Roosevelt had a similar complaint, writing that for others at university "the tendency was to treat as not serious, as unscientific, any kind of work that was not carried on with laborious minuteness in the laboratory." He abandoned "all thought of becoming a scientist." Yet he went on to become one of the most important naturalists in history, fathering the U.S. national park system and thereby creating a precedent for other countries around the globe. But my inability to keep my head above water in academics worried me: how then would I continue to work and make an impact in the Amazon, in a field where everyone of consequence seemed to have a PhD? In this way, the future was an unanswered question.

Despite my run-ins with biology professors and defeats in ecological statistics classes, Ramapo College had been a good choice in that there was a crew of vibrant, dedicated professors inhabiting the environmental science department. Through subjects like forest resources, water resources, biology, and ecology, I found myself, for the first time in my life, actually getting something out of education. This was partly because the people teaching me were, in essence, grown-up versions of me. Of particular note, the professor who taught a course called Ecology Economics and Ethics would influence my life in big ways.

Trent Schroyer had spent his teenage years as a fur trapper in the mountains of western Maryland but eventually realized that animal populations were declining. So he hung up his snares and turned scholar, conservationist, author. When I met him he was in his seventies but still furiously turning out books and organizing international protests to support ecological and social change and bring down unregulated capitalism. He was a rebel if I ever saw one. A direct man, in today's world of political correctness and delicate sensibilities he was viewed by many as a tyrant. To me he was a brilliant, friendly breath of fresh air. I understood the reason for the gruffness he showed at times: it was as though he had crammed his mind with so much information about the problems facing our planet, and the literature on how to fix them, that he had altogether lost his tolerance for bullshit.

The content of the class focused largely on understanding the forces behind the destruction of nature and how it impoverishes humans. The commons, as he called it, was a new way of thinking about the natural world; the idea was that water, air, forests, and even animals are all "commons," which don't belong to anyone—they should come free and guaranteed, a universal endowment, for all life on earth. They are the inheritance of every living thing. The problem, as we learned, is that today many of the things that should be inherent, abundant, free, and pure are disappearing: the oceans are being overfished, the forests are all being shaved, air and water are polluted and privatized by corporations, land is bought up, wildlife exterminated.

For the first time in my academic career, my experiences and interest in the real world mattered inside a classroom. Trent was interested in the Amazon as a woodsman and as a scholar of the commons, and we spent hours talking together. He was a man of great warmth and depth, a fact obscured to almost everyone by his brash brilliance. Somehow, despite the decades that separated us, there was a mutual fraternal recognition. As that semester went on, we spent an increasing amount of time discussing our mountain-man adventures and environmental theory. It was on one of these days at a café between classes that Trent first brought up the prospect of India.

He explained that he had created, and for the last several years captained, a study-abroad group to India, and he felt that it would be an important learning experience for me. "This is a place you need to go," he said, looking at me frankly. "The Amazon is intact in your area at this point, and that is great, but what is going to happen ten, twenty, fifty years down the line? You need to learn what is coming, so that you know how to protect it. In India it's all happening now. The time is going to come when you are going to have to protect this place you love."

Unafraid to put me on the spot, Trent persisted. "Are you a thrill-seeking adventurer, purely after your own enjoyment? Or do you feel that you might possibly owe something to the jungle, have some responsibility for it?" I could see what he was saying but still was not thrilled about the idea of going to a place with more than a billion people. "Besides," Trent added, strategically, "how could you not want to go to the country that has jungles filled with elephants, and more tigers than anywhere else on the planet?"

Several weeks later, I was on a plane to Bangalore, India.

Trent ran the semester-long program out of a place called Fireflies, an ashram; it was an aesthetically brilliant spiritual/learning center surrounded by tropical foliage and green architecture. There were birds, snakes, frogs, and other wildlife sneaking around the campus and I instantly liked the place. It was a serene contrast to the sprawling city nearby.

For any Westerner, arriving in India can be shocking. I remember walking out of the airport, and the first moments of the madhouse: bustling streets filled with thousands of people, goats, carts, cows, lorries, wagons, horses, dogs, cats, vendors, buses, rickshaws, motorcycles. Once, on a road full of traffic in the state of Kerala, I saw a tremendous bull elephant, as tall as the buses, with several men on its back and a load of timber on its tusks. The smell of spices, burning eucalyptus, and an entirely alternate reality from anything familiar—it is pure, beautiful insanity. As per the cliché of Americans who travel to India, it was a transformative experience. I quickly fell in love with the place that would eventually become my home. But at the same time, as a naturalist, and just as Trent had predicted, I was frightened by it.

Farmland surrounded the ashram, occasionally interrupted by tufts of forest, and I would sprint from class each day into these in search of giants. More than one person told me stories of elephants coming over the fields to the ashram to soak in the lake and pillage crops. I spent virtually all my free time roaming in search of even a single pachyderm. It was weeks before a friend at the ashram, a young guy named Rajesh, asked what I was doing for so many hours out by myself, and I told him. Then he laughed. "The elephants aren't coming here anymore, boss," he said, smiling and wobbling his head. "Years back they were coming every day, you know, but that is because"—he pointed toward the horizon—"that used to be forest. They would be coming through the trees to visit the lake each night, but the forest is gone now. Long time ago."

To Rajesh, elephants coming or not coming was about as important a matter as the kitchen serving masala tea or cinnamon tea. To me it was very different. For weeks the land had seemed vibrant with possibility, excitement, and potential—just at the idea of elephants. Each new bend in a brook was an adventure when a herd of elephants could be around the turn. But hearing that they had been chased away and run out more than a decade ago changed everything. The landscape seemed mute and uninteresting. After that I had less reason to go out. This learning curve was a major theme in my early days in India. So I resigned myself to spending much of my time drinking chai and reading books, hanging out with Rajesh and the other workers of the ashram, playing highly competitive games of carom late into the night.

Yet still the theme of India for me was the tiger trail. I had come to India imagining seeing a tiger the way I would stalk an animal in the Amazon. I had read enough to know that if you can find a deer carcass, a tiger kill, sooner or later you'll see the tiger. That, I thought, couldn't be very difficult. I would hike through the forest until I found a deer carcass, then stake out the site; all alone in the woods, I would watch the greatest predator on our planet feast. But it was far from that easy.

Just a few hundred years ago tigers ranged from Eastern Europe and the Middle East and down through Asia and Indonesia, all the way up to the frigid eastern Russian peninsula of Kamchatka. They were the ruling terrestrial predator of our planet. They are the genetic pinnacle of millions of years of evolution—the ultimate predator. From the steaming jungles of Sumatra to the snowy wilderness of the Russian far east, they are the apex predator—the latter of these species, the Amur tiger, even dominates the brown bears and wolves that share their range. Yet humans have historically been at odds with the fire cats, mercilessly hunting them, poisoning them, removing their habitat and prey species, and sometimes deliberately exterminating them. Many have speculated that it is our similarity to tigers rather than our differences that is the true impetus for the ancient conflict. Today the species is devastatingly close to extinction. More than 95 percent of the tigers that used to inhabit the earth are gone; most of those remaining are found in India. It is a species in triage.

I searched for tigers in India the way I had for anacondas in the Amazon, asking people, searching for insights into where to find them. But I was following ghosts. Old farmers would tell stories of the tiger that had eaten a cow, or been seen on the edge of the forest when they were young. Their eyes would glow with the memory of the wondrous creatures. Products in stores used the tiger's image, vendors sold trinket statues of them, and houses often had at least one ornate wall painting that bore the image of a tiger. But, despite their omnipresence in people's minds, the stories always had the same ending: that was then, this is now. The forests had been cleared, or occupied by the homeless and subsequently emptied of wildlife. In the vicinity of Bangalore no one had seen a tiger in decades.

They were gone. My quest for tigers taught me that India has changed at dizzying speed in the last fifty years, socially, economically, and environmentally. The India in the minds of the elderly and the India before my own eyes were different worlds.

With mounting dread I spent months researching, making phone calls, learning. Every lead directed me to a national park or reserve set up specifically for tigers, where tiger tourism was run by way of organized jeep travel, and all other entry was illegal. In years to come I would learn about the importance of these reserves and come to understand that without them, and the hard work Indians do to maintain them, tigers would face an even greater magnitude of doom. Against the tsunami of development and population growth, dedicated and heroic scientists, photographers, naturalists, politicians, park guards, and reporters had been devoting their lives to protecting the striped cats long before I was born. Nevertheless, coming to terms with the fact that I would not be able to simply strike out in search of the cats in "the wild" was infinitely frustrating.

In my first trip to India, as a young, dumb student, without many connections, I spent most of my time reading about tigers instead of actually hiking through forests. India's tigers are contained in roughly fifty designated tiger reserves across the country. Along with the tigers, of course, one finds in many places elephants, leopards, wild dogs, Indian rhino, deer, boar, hundreds of bird, reptile, amphibian, and fish species—all protected under the name of the "king," the tiger.

Just like at Yellowstone National Park in the United States, there is a front-country and backcountry philosophy to Indian tiger reserves. Many parks incorporate native tribal villages, roads, guard stations, and even grazing areas for livestock in the front-country areas. Yet each, or most, have a backcountry, which in India is called a "core zone," an inviolate area where the idea is to have no human activity at all (or as little as possible). Like any system, it is not without its flaws, but as I dived into the literature of Indian wildlife conservation, the core zones struck me as ingenious—allowing the wildlife space apart from humans, to breed, feed, and go about their lives in peace.

Yet to my great despair, these core areas were completely off-limits, and getting authorization to enter them was a monumental feat. In years to come, as I became more familiar with the Indian system, I would come to value and treasure this fact of inaccessibility, for its ability to keep the animals safe—but during that semester abroad it was frustrating. My dream had been to move through forests alone, tracking, learning—not brought through on an official, expensive jeep packed with tourists. I had read enough to know that when a tiger or other notable animal is spotted, there can be dozens of jeeps bearing hundreds of tourists, all snapping photos, all violating the peace and privacy of the very animal they are marveling at.

No, I wanted to be walking silently through a forest and see a tiger, a moment I fantasized about so often it became almost an obsession. But as a newcomer to the country I was floored by how few human-free spaces there were. In the past fifty years the explosion of humanity in India had consumed every scrap of habitat that was not protected by law. Every last chital deer, every last boar in many places had been hunted out, thereby removing the tiger's prey base. The tigers were gone and the trees fewer, the herds of elephants more distant. At the time I first traveled to India, a national debate was taking place over who had the rights to live in India's remaining forests: tribal people or tigers. In reality, both groups had been given the short end of the stick by the greater society, which had cut down so much of the forest that once stood. The battle over who is entitled to the patches that remain, tigers (and the other wildlife encompassed in their domain) or tribal communities, is ongoing and fierce.

Archival photos in India show the result of British elephant-back hunts, where proud colonial aristocrats pose besides dozens of tiger carcasses. As India's population grew, the conflict between humans and tigers increased, and in many places hunters were hired to eradicate the cats. This deliberate culling, combined with the habitat destruction of a rapidly developing world, saw India's tigers reduced from 40,000 to only 1,800 in the span of a single century. Think about that for a moment. Today most tigers in India exist within the confines of specially designated, guarded areas and national parks, and continue to do so only because of dedicated people who fight the myriad forces that are forever threatening to complete the extermination of the great cats.

What I saw as a student, new to the Indian reality, was a tragedy. Despite the national parks, and the incredible work of local and international conservationists who work to ensure the survival of the species there, only a scrap of the past remains, a stump of a once-glorious tree of life. And as always, it is crucial to reiterate that under the umbrella of tigers fall thousands of other species that are less iconic, yet equally as deserving of protection. The subject encompasses entire biomes. India's population has strained resources to such an extent that conflicts over land, water, forests, minerals, and indeed all natural resources have reached the breaking point.

During the semester in Bangalore a visiting lecturer asked: "Why do we need the tiger? Why do we need elephants?

Do we need them at all? If the farmers don't want them on their land, and the developers find them a nuisance, and as long as there are tigers in parks so that tourists can gawk at them, what use is it to us to have tigers running around in the wild? Let the forests stand and regulate climate, and rainfall and what have you, let them be there. But tigers and elephants? Maybe it is of no use and they should be extinct? Why would we want elephants to come and eat our crops? It would be better if they were gone. Life is difficult enough already; how can we spend time trying to protect animals when humans are suffering?" There is perhaps nothing so tragic as the idea that the vanishing of species is a logical part of human progress.

I did not get to see a tiger on my first trip to India. Instead I spent countless hours learning about the complexities facing them. Yet amid all the studying, I did find some time for fun. I spent time playing carom with the staff of the ashram, hanging out with the other students, and occasionally traveling a bit.

One day, near Mysore, I woke up early on a free day during a class trip and made for Nagarhole National Park, a dry-forest tiger sanctuary. Though my professors had told me not to go, I was not about to pass up my only chance, and at the time had no idea that I was committing a grave offense against the forest department, not to mention entering a forest inhabited by elephants, tigers, leopards, bears, and numerous venomous snakes.

It was within Nagarhole, at last, that I got a first look at Indian wildlife. On a single day's hike I spotted peacocks, a sloth bear, chital deer, herds of titanic guar (Indian bison), and numerous other animals. Being in a forest that definitely held tigers supercharged every step; I was on cloud nine. Though it took considerable effort to find out where to go, and a long hike around checkpoints to get in, the forest in India was overwhelmingly beautiful.

Toward the end of a long day, I rock-hopped silently through a shallow gorge as the slanted orange rays of the afternoon sun lit up the lime-green forest. A troop of black-faced langurs with luxuriant long white tails observed my passage from the canopy. The stream I followed slithered between some large boulders, and as I leapt from one to the other I noticed that there were huge circular depressions in the mud, and large piles of fresh, grassy droppings. With mounting excitement I took great care to remain silent, and stalked onward. When I rounded a bend I met a sight as terrifying as it was wonderful: a cow elephant.

To see an elephant in a forest, just a few dozen feet from where you stand, alone, is an experience I cannot do justice to with words. Suffice it to say that a wild elephant that has lived in proximity to angry farmers and other human conflict can be among the most dangerous animals on earth. But there was nothing malignant in either her posture or expression. Her brown, heavily lashed eyes looked into mine as her trunk calmly worked below. The mud-caked giant barely looked real, towering more than nine feet tall. She looked more like an animated cement statue than flesh and blood. Amid the orange beams and delicate lime foliage of the wood, she was a vision.

It was unnerving how intimately she seemed to calculate and quantify me. She knew I meant no harm but was small enough to kill if necessary. For several minutes I remained still, as she gathered vegetation in her trunk and sniffed about. But when she saw that I was not leaving, she cautiously, almost silently, melted away into the brush. I was left alone in the forest with my heart jackhammering and hands shaking.

In just ten hours in Nagarhole I saw a world I had never known existed; it was a glimpse of what India's Karnataka landscape had once held here. But this was not wild nature. Nagarhole was filled with roads, tribal villages, guard posts, and various degrees of human interference that every species was forced to contend with. My goal in India was to find the Indian wild, if such a thing still existed.

Over the months I thought about the Amazon almost incessantly. Being away from it allowed me to view it with a new perspective. I thought of the secluded research station, and the anaconda expedition with JJ, Chito, and Pico. More and more I began fantasizing about one day returning all the way up to the point where we had had to turn back, and going on. It became a recurring daydream, a world in which I spent more and more time. The more time passed, the stronger the urge grew, and gradually I concluded that at some point in my life, I would escape the world and journey into the deepest parts of the Amazon alone. Somehow, it seemed like I was obligated to do so.

When I wasn't studying or exploring, my time at the ashram was peaceful, and highlighted by new and interesting relationships. And the fact that there were plenty of snakes around campus. By the first week I had become established as the nutjob-who-catches-snakes, in a countryside where virtually every person fears and kills them. Trent had pointed out that beneath the many termite mounds around the ashram, some as tall as a man, there were holes where cobras often lived. People of the Indian countryside had great respect and fear for these lairs and the snakes that lived within them, especially cobras. One man told me that if you see a cobra and don't kill it, it will slither into your bed during the night and kill you. He killed every snake he saw. Wherever you go in the world, snakes are among the most gravely misunderstood animals, often deeply tied to superstition and fear.

Though I have never been superstitious, it is a fact that snakes have always been harbingers of great events in my life. The first snake I ever saw, in third grade, was a brilliant blue garter snake. I spotted it while walking with a teacher through a patch of woods and instantly wanted to hold it and marvel at it. For young me, it was a momentous and wondrous moment in life. My teacher, knowing that there was no way of stopping me, scared it away, and then left. Distraught, with tears running down my cheeks, I remained in the woods searching for the snake. Noel came by just then and did his best to help me look for the snake, overturning rocks and logs—and thus began a lifelong friendship. On my first trip on the Las Piedras, JJ and I jumped into the river to go after that huge whip snake, an event that jump-started a friendship that led to us unlocking the Amazon. Had I not met JJ, or connected with him as I did, there would have been no Lulu, no Pico, and no anacondas.

Throughout the semester in India I caught all manner of snakes, from yellow rat snakes to vipers and even cobras. But it was a small checkered keel-back, an unimpressive water snake, that heralded the most important intersection of fates in my life, and was the result of a friendship forged in a tragic moment.

Over the semester Rajesh, the grounds manager at the ashram, who had been so curious about my elephant searches, became a close friend. He found my snake love puzzling and terrifying, and would often run screaming when I'd burst in with a new capture. He preferred to spend peaceful evenings teaching me the rules and skills of the game of carom, which I came to love. We spent a lot of time together, and I returned his tutelage by helping him practice his broken Indian English.

But one day in the computer lab, researching tigers and elephants, I heard a cry that sounded like Rajesh. Then I heard other voices. Instantly the hackles on my neck rose. Most of the students on the study abroad program had gone into the city for the day to shop and party, leaving just me and one other student working. We looked at each other as the din of terror-stricken voices grew in magnitude: something was happening.

Rushing from the tranquil computer lab and down the stairs, we ran toward the noise. What we found was confusing chaos. A dozen men from the surrounding villages were drunk and furious, carrying rakes and poles and other implements of destruction. They were shouting and frantic, with blood-lust written all over their sweating faces. The workers of the ashram, many of whom I had become close to in the previous months, were in a state of terror.

They had sought safety from the mob in a caged patio: workers inside and assailants outside. The village men swung their weapons against the cage; they wanted in. A woman stood at the gate, desperately imploring them to calm down, but they were rabid with rage. One of the villagers kicked the gate and sent the woman backward into the stone wall, where her skull hit the unyielding brick and she fell limp to the floor. The other women within the patio cage shrieked in terror and for the first time I noticed my friend Rajesh on the floor in the corner. His face was bloody and swollen from blows, and his clothes were torn. Several others were shielding him with their own bodies from the oncoming attack. Though I couldn't understand the words that were being yelled by attackers or victims, it was suddenly very clear what was going on: the mob was trying to kill Rajesh. I knew I would have to wait to find out just why they wanted him dead (although in the end, I never did find out what caused the conflict).

As the gate swung open, a young and very skinny man took over where the woman had fallen, straddling her limp body. But the mob was seething, shouting, and their viciousness was now concentrated wholly on the young man holding the gate. They landed repeated blows on him through the bars, and his courage made my chest shiver. In the dilated, adrenaline-fueled super-reality of the moment, I suddenly wondered if I was about to see a person beaten to death before my eyes. Now they were slashing at him with a machete. I turned to see my student friend from the computer lab, but he had vanished. The mob assaulted the gate, slashing and cursing and spitting. I knew I had to do something but was kept still by the reality of more than ten men with weapons against one kid with none. It was terrifying.

However, as I watched the struggle at the gate door an idea surfaced. Practically flying, I sped to my room and grabbed a large padlock and my camera tripod, and a large bowie knife, which I stuffed into my pants. In a few moments I was back on the scene, where now even more people had gathered. Another ashram worker, husband to one of the women in the patio, had tried to reason with the mob, and when I arrived they rained punches and kicks onto his limp body on the ground. The young guy inside the gate was barely hanging on. Rajesh was bleeding badly within, and a woman shrieked through the bars in desperation as the mob beat her husband on the ground where he lay. Still, though, the focus of the mob was on getting to Rajesh.

In panic, I jumped into the fray of bodies and thrust my arm through the bars, meeting the gaze of the bloodied young man who was holding the gate. He had taken several blows, but in that moment we locked eyes as friends—albeit ones who could not speak a single word of the other's language—and I handed him the lock. I put my shoulder into the crowd and took a rattling blow to the head while buying my friend the crucial moment he needed to close the gate fully and slip the padlock through the hole. Once he had done it, I retreated. The mob was furious and began throwing stones and thrashing the gate with their farm tools, demanding Rajesh. As their assault intensified, it suddenly seemed that they would rip the entire patio cage to pieces.

At this crucial moment a new figure arrived on the scene. He was well over six feet tall, a villager. But he was not part of the mob. He approached the savage scene wearing his skirt-like *lungi* and button-down shirt, his impressive pillar-legs set firm on the earth. He roared like a bear and threw one of the mob clear through the air, over the heads of his fellow attackers. Suddenly they all paused, shouting new threats toward the man. He was a giant. The biggest Indian I have ever seen. His stern brow shadowed eyes of black fury above a heavy black mustache. He faced the drunken mob with a quiet promise of pain, as his hands patiently tightened the knot of his lungi.

Like a pack of dogs, the mob turned on the man, shouting and brandishing their weapons, but hesitating to attack. He stood strong, knees bent and powerful arms hanging ready at his sides. Was he going to take on all of them? Onlookers stood transfixed. In a moment of thoughtless support, I rushed to his side, with my camera tripod poised over my shoulder like a baseball bat: two of us against twenty men. For a palpable moment all-out war seemed imminent. But as we stood our ground, others joined us. With each additional person on defense, the mob, like all cowardly mobs faced with real opposition, began to lose its enthusiasm.

The day ended without any deaths. The mob was persuaded off the grounds and we spent the day cleaning up Rajesh, and standing watch in case they mounted another attack. The woman who had been knocked out came to, and we all had chai and nursed the wounded in the kitchen. The young man who had held the gate and I sat together for many hours, drinking chai in silent recognition of what had passed. In the days that followed, peace eventually resumed at the ashram.

The strong man in the lungi with the impressive legs and heavy mustache vanished before anyone had a chance to thank him. However, weeks later I saw him once more, riding his bicycle. I waved to him and for a moment he regarded me with a scornfully stern gaze before recognition dawned on his face. He nodded, with a wry smirk of acknowledgment that we had stood together in that dire moment. Not a single word passed between us.

After that Rajesh and I were brothers. We spent even more time playing carom and practicing English. He began calling me whenever someone saw a snake, so that I could capture and relocate it, instead of it being killed. It was a good system. I repaid his many gifts by helping wherever I could. He was in charge of the festivities that surrounded the ashram's annual music festival, and I helped him along with the other students to prepare. We spent weeks decorating, organizing, and planning.

On the night of the show, thousands of people flocked in. It took place in a stunning stone amphitheater that descended in large pews to a stage over which an old banyan tree grew. It was a beautiful spot. Everyone at Fireflies prepared night and day for weeks beforehand, and when the night finally came it was well worth the effort. Bands from all over India played the music of their heritage, ranging from the ancient to traditional to fusion Indian rock. The audience was like what you would see at an Olympic opening ceremony in terms of diversity, with people from all over the world. The energy was explosive. It was a festive night, and as I helped usher bands from their rooms onto the stage I kept sneaking shots or tokes from friends; by midnight I was pretty buzzed.

Amid the haze of light and sound Rajesh came running and grabbed me by the wrist. "There is a snake by the stage, boss, you have to come!" Instantly I sprinted after him. We both knew that if the locals got there first, they would surely kill the snake. As I arrived near the stage people were recoiled in fear, staring at a hole among the roots of a tree. "Where is it?" I asked, and several helpful onlookers informed me that the snake had vanished into the hole. The snake was safe, but people were still scared.

Just as I was reassuring people that there was no way the snake would come out again, a girl skidded to a halt directly in front of me. My eyes widened when I saw her black hair and big eyes; she was stunning. I had been scanning the crowd all night, admiring the abundant female beauty, but this was different. We all go through life with an image of what we desire in another, emotionally, physically, etc., and frequently copy and paste various traits into an unfocused collage of who we hope is out there, and I instantly recognized the girl before me as the focused manifestation of mine. I was breathless. "Where is the snake?" she demanded with melodic authority. I smiled and countered with "Why would you like to know?"

She gave me a vexed grimace. "Because in my country they will kill a snake if they get the chance. I have to catch it and get it out of here. Now, where is it?" I was mystified. Who was this beautiful girl who caught snakes? I explained that it had gone into the hole, and pointed out where. She dropped to her knees and peered in. If her physical beauty was the first punch I received, her genuine concern for the safety of a small creature was the finishing blow. Now I was swimming. Who was this girl?

"Hey, look," I said, "I work with snakes professionally. It won't come out again tonight; it should be fine." She was standing now but still gazing toward the hole to be sure.

"You work with snakes?" she asked.

"Yeah, anacondas, bushmasters—all different stuff—in the Amazon." I was hoping that she would have been more impressed than she was.

"Cool. This was probably a cobra. Have you ever worked with cobras? A few months ago," she continued without pausing, "I was in Rajasthan and we were on elephant back in the high grass and we saw a king cobra; it was incredible." She could see my questioning look and explained: "We were on a class trip and went tracking rhinos, which has to be done on elephant back. Actually, we found one and it charged us. Pretty scary! We came so close to getting . . ." As she spoke, the exotic orange glow from the stage and ancient foliage of the banyan's branches illuminated her features. I tried not to stare at her, but I was reeling from the shock of what I saw and heard. I remember at one point during the night seeing Rajesh smiling and wobbling his head at the sight of so many sparks resulting from his snake call.

Though I talked to this girl as much as I could during the

long night, I remained mostly occupied with various jobs, and by morning we politely exchanged email addresses and parted ways. In the months that followed, I wrote to her many times, but I never got a single message back. I tried to convince myself that my mind was embellishing memories, but in my mind the girl at the music festival had become *the* girl.

Despite thinking about her all the time, I tried hard to shake it off. The foundation, rule number one, for me, had always been to remain free; to focus on exploration and wildlife, and hone skills that would allow me to push the boundaries of exploration in the Amazon. At the time, the idea of getting tied down by some girl was detestable—especially one that lived all the way across the globe in India! With determination, I coached myself into letting it go.

During the semester I had become good friends with a guy named Ananda, whose father ran the ashram. Ananda and I shared a fascination for snakes and other wildlife. Together we planned to travel when our classes were done to a rainforest reserve in the Western Ghat Mountains. There, Ananda promised, were tigers and elephants, as well as snakes and many other creatures.

Just two weeks before I was to meet him for the journey, he phoned to ask if a good friend of his could join us. "She said she met you at the music festival, something about a snake? Her name is Gowri." How it happened I still don't know, and I won't dare begin to calculate the odds, but somehow the one close friend I had made in India was a classmate and friend of the girl haunting my memory.

We planned to meet in two weeks and head out to the jungle. In the interim I could not help but be excited, despite my efforts to the contrary. We had agreed to a rendezvous point where I would meet up with Ananda and Gowri. When the day finally came, Ananda waved as his rickshaw approached. When he reached me he hopped out and gave me a hug, and that's when I saw Gowri inside, beaming. Without exaggeration, I can say that when our eyes met this time, we both knew.

The three of us spent a magical week together amid the unfathomable beauty of the Western Ghats rainforest. Though the area is not in any way comparable in size to the Amazon, I found myself in similarly wondrous rainforest. Hornbills and monkeys called from the canopy, and herds of elephants left tracks in the streams—even a few tigers made their living in the verdant rainforest. Rich, flowering foliage supported limitless wonders, from Malabar gliding frogs to king cobras to stunning flowers. In an orchid garden, Gowri and Ananda and I would marvel at pit vipers; once we found a stick insect as thick as a man's finger and ten inches long.

With each passing minute it became unbearably obvious that there was a powerful magnetism between Gowri and me. We climbed strangler figs, caught snakes, and adventured together. She was as at home in the jungle as I was. I had never been so lost in someone, so completely at ease. Being with Gowri was a different plane of reality; there simply weren't enough hours in the day for us to interact. She had an honest and unassuming, almost childlike enthusiasm, and a way of existing in the moment that was unlike anyone I had encountered. In the refreshing glow of her energy I felt like I had for the first time found a member of my own species.

By the time the week had ended I was desperate to continue my relationship with this incredible girl and see where our friendship would go. But I was flying back to the United States and down to the Amazon in just two days.

Leaving the jungle with Ananda and Gowri, we boarded a sleeper bus back to Bangalore. Over the course of a night spent watching the Indian countryside pass by while others slept, we made the most of our last moments together. To make a long story short, after much whispering, positioning, and gradually drawing closer, we took advantage of the only chance we would ever get and kissed—kissed like the world was ending, which in a way it was.

It was as exciting as it was tragic. There was no longer any denying the enormity of what had swept us up. Moments after it happened Gowri began to cry, and we both confessed to holding back the entire week because of the logistical wall we were up against. I told her that all I knew for certain was that every molecule in my body was aware that this was not the kind of connection that happens twice in a lifetime. Still, I was leaving in just a few hours for the other side of the world. In all likelihood, life would go on, time would pass, and we would never see each other again.

By morning we arrived in Bangalore and got breakfast and the world felt new, somehow on fire. It was wonderful and horrible. Once again we were carried by luck. I had gone back to the ashram to pack and promised Gowri that I would meet her in the city before flying out the next morning. But with my bags ready to go, mere hours away from leaving, I noticed my passport was missing. After a few frantic hours of searching I realized that weeks earlier I had left it with a friend while I went swimming. That friend was now a thousand miles away. There was no way I was flying out tomorrow.

I called Gowri barely able to contain my excitement. At the twenty-fifth hour we had been saved: the passport would take more than a week to reach Bangalore. In that time Gowri and

I were inseparable, electrified by our new relationship and the knowledge that the time we had together was a rare stroke of luck, a gift from the universe.

The last week in India I had nowhere to stay, but after we had dinner with Gowri's family, her parents were generous enough to put me up in their guest room. Gowri's room, which she shared with her sister, was much the same as my own room at home, a museum of natural artifacts: butterflies, skulls, fossils, and leaf skeletons retrieved from years of adventures. She even kept a box of snake skins that she had collected, just like I did at home. Being in her home, I found she had a wonderful, wild, and very endearing family; it didn't hurt that they cooked some of the best food I had ever eaten.

Each night we'd sneak up onto the roof with a few blankets and stay up all night. On the last night, wrapped in a sleeping bag together, I remember praying that time would stand still. But the Muslim call to prayer echoed over the city as morning crept into the east. Whispering, we came to the absurd conclusion that we had to try to stay together. I promised to one day show her the Amazon, and that we would travel the world together, but most important, I promised that I would be back to India soon. Even as the words came out of my mouth, I knew that the likelihood of a seventeen-year-old girl from India and a twenty-one-year-old guy from America maintaining a relationship was slim. But each time I tried to be "realistic" and discuss what would inevitably happen as time and space separated us, she stared hope into me with wide, determined eyes that squashed the doubt I felt. There was no doubt in her. What could I do? When life throws a gorgeous, energetic, snake-catching, animal-loving girl into your arms, you don't just walk away. I knew that if the relationship was allowed to fade, I would never experience this kind of connection again. And so, kissing the two freckles on the left side of her nose and crunching her in my arms, I promised her it would work.

I left India, spinning from the nascent relationship with Gowri and the cosmic injustice of being torn apart. Yet within I also heard the call that had been nagging me for months: *get back to the Amazon*. During the recent months the bliss of my ignorance had been obliterated by what I had learned of the world, learned of tigers. I desperately needed the unspoiled green solace of the Madre de Dios.

With a volunteer group planned far in advance, I had only five days in New Jersey before I repacked and headed for Peru.

12

The Beached-Whale Paradox

Until my ghastly tale is told, this heart within me burns.
—Samuel Taylor Coleridge, *Rime of the Ancient Mariner*

I once made a decision so ludicrously rash it nearly killed me, and by all rights should have. It occurred during my first solo excursion and resulted from a confluence of events that blindsided the Las Piedras. It was soon after my return from India, when I was old enough to grasp the scope of what was unfolding around me but young enough to lose my mind as a result. The psychotic adventure it sparked would take my relationship with the Amazon to places I never dreamed possible. It started with some very bad news.

It occurred while my world was still an emotional blur after leaving Gowri. JJ met me at the airport and hugged me for a long time and told me I'd been away far too long. He also told me that he and Emma had sold the station. It was no longer ours. I tried to figure out what this meant for my future, but there was no way to know. All JJ could tell me was that the buyer was a larger organization from the United States, and the paperwork was being put together. He said this could be the last trip we would make to the station. The question of how my life in the Amazon would continue, or if it would continue, saturated my consciousness.

Working to suppress countless questions, I did my best to engage the volunteers we brought by spotting wildlife and explaining interesting facts about what we saw, and a few groups passed uneventfully. I tried to savor the station, the forest, and as always kept an eye out for signs of Lulu, but I did not feel at ease. I was twisted by the uncertainty of the future. It was even worse watching JJ suffer as Emma prepared to leave permanently for the United Kingdom with Joseph.

On a day in between visiting groups, JJ and I were bringing supplies up to the station. As always, as we rode, I scanned the vines and towering trees—the massive green walls that fortressed the entirety of the river—and soaked in the visuals of Las Piedras. Rounding a bend, I smelled smoke and turned to JJ. "Who's burning stuff way out here?" I asked. From his expression, it was clear that he had not told me something.

The scene entered my mind like black poison. Interrupting the previously endless walls of green was a gaping hole in the jungle, where the forest had been completely burned to the ground. Smoke rose in twisting columns as the air shimmered above the inferno. Ash fell like snow onto the river. Everything was gone.

Suppressing tears, we surveyed the burned wreckage. Thousand-year-old trees lay slain and charred across the ground. Among them were the mangled remains of palms, vines, amputated buttresses, and other vegetation. Explosions shocked the air as large bamboo chambers burst, spitting ash and flame aloft. Hundreds of years of photosynthetic growth, thousands of species, millions of years of evolution, razed. JJ and I stood in devastated shock at the wreckage that had once been forest.

Yet the destruction was nothing more than a sample, just a finger of a much larger demon. The clearing was part of a new road that had stemmed from the trans-Amazonian highway, or BR-361, possibly the most environmentally devastating single project in the history of the world. It was announced in 1970 by the Brazilian government as a strategy to integrate the Amazon with the rest of the country. The plan was to slice a network of pioneer roads into the unbroken forest that would provide access to the vast mineral, timber, and agricultural resources there. The two-thousand-mile Rodovica Transmazônica was to be the backbone of the proposed web of transportation that would essentially serve to open up and tame the wild west of Brazil. Funded by the World Bank, it was part of Brazil's Program of National Integration. In the 1970s, Amazonia dominated half the Brazilian territory and held only 4 percent of its population. The plan was to provide incentives to many of the poor living in the east to migrate west and settle the Amazon.

In a time when Third World countries were expected to follow the Western model of development, large swaths of jungle were considered undeveloped. In the 1970s organizations such as the World Bank and International Monetary Fund were more than willing to provide the financial assistance that poor ex-colonies needed to pursue this model, and so $440 million was granted to the Brazilian National Highway Department.

Much like the United States at the time of its inception, Brazil's population is concentrated mostly in the eastern region of the country, while the west remains a mostly untapped wilderness of Amazonia. The desire to push westward into the so-far inaccessible 60 percent of their country informed the Brazilian consciousness, and still does.

By 1975 the entire unpaved road had been bulldozed through, a laceration that ran across the entire southern face of the Amazon biome. As a result, dozens of uncontacted and semi-contacted indigenous tribes were suddenly in the path of *development*, and under assault from germs, development, and migrant farmers in a lawless wilderness. Farmers poured into the Amazon behind the heavy machinery, striking out in right angles from the road and settling vast tracts of jungle by turning them to ash. From the air, the scars can be seen today. Scientists call it the "fish-bone effect" when a single road is blazed and then many others spring from it. From above, the roads seem to draw a fish skeleton. Even as a child, after my visit to the Bronx Zoo, I knew that once a road is made, rainforests disappear. The relationship is unfailingly direct. The trans-Amazonian highway shot thousands of fish-bone roads into the jungle, opening up previously inviolate habitat to the degradation of hunting, farming, gold mining, etc. The megahighway was a terrible blow to the Amazon, but because it remained unpaved, its effects remained only a fraction of what they could be.

At the start of the 1970s the world entered a period of significant discovery about the state of the environment, and a greening of international politics took place on a large scale. By the 1980s, under mounting pressure from an environmentally aware public, the U.S. government conducted a hearing on the environmental impacts of the World Bank's loans to Third World countries—which an increasing number of people believed to be more destructive than good. In the past the World Bank had been considered something of a hero for helping Third World countries to "develop," but it was becoming clear that giving vast sums of money to poor governments often further impoverished more than it empowered; millions of people were displaced and vast environmental damage was done by hydroelectric dams and highways.

For the World Bank's then-president, Barber Conable, this pressure was real, and a sweeping reorganization of the bank took place. It became more strict in its lending and required more detailed environmental impact statements before approving loans. The bank stopped its funding of the Brazilian highway system in 1985, effectively halting all construction on the road, which would have linked Brazil to the Peruvian border near Puerto Maldonado and eventually run over the Andes to the Pacific Ocean, opening up the Amazon to the Asian timber market by a direct land route to the sea for the first time in history.

Furious, the Brazilian government tried obtaining further funds from the Inter-American Bank to pave the trans-Amazon highway. Environmentalists and the people of Amazonia in the state of Acre and other areas reacted with urgency, flying iconic local activist Chico Mendez to the United States, where he addressed the World Bank and American organizations, urging them to deny funding. The loan, thankfully, was never approved, though Mendez would later be assassinated in retaliation for his efforts to protect the forest and its people.

For decades the trans-Amazonian highway lay in disrepair, a tragedy for conservation and a failure of development. The unpaved road was overgrown and functional only in certain places. Each rainy season the torrential downpour would flood the land, causing massive erosion and further damage to the highway. Because of difficulty navigating, the road was expensive to travel on and the jungle began to reclaim what had been taken from it.

This scar across the continent was broken in only one place: the Madre de Dios River, where the only way to cross was to drive trucks and cars onto boat ferries, which was costly and slow. In order to complete the highway's planned course from Brazil and over the Andes to the Pacific Ocean, the architects of the highway had planned a bridge that would cross the Madre de Dios River itself, directly across from Puerto Maldonado. When I first visited the region, there were two large concrete pillars standing in the river alone and unfinished, fossils of an extinct monster. Emma explained they had been built to support a bridge—and warned that its completion, and the paving of the road, if it ever happened, would mean the end of many things. She spoke of the past and future with stony eyes, weary with the desperate inner prayer that what she spoke of would never come to fruition. She knew full well that in conservation the victories are temporary; it is the losses that are final.

The burnt forest and crude road were just an hour downstream from the station. What had for all of history been deep and inaccessible jungle was now exposed, violated. What I saw robbed me of sleep for weeks, because I knew this was only the beginning. A road meant that settlers were coming. In years to come a town would sprout up, more and more forest would be burned, and hunters would penetrate for miles to either side of the scar in the forest. Then the logging would start.

When JJ and I returned to Puerto Maldonado, things only worsened. Like being in a nightmare that would not end, I stood on the high banks of the Madre de Dios River staring at the large orange cables and tension lines that were being stretched across the river. The bridge was under construction as the highway, the giant that had lain dormant for decades, was waking once again. Suddenly it felt like the Madre de Dios was a hostage tied to the tracks, awaiting the inevitable iron force that would tear it apart. "Progress," they say, must march on.

Scientists all over the globe agree that we are in the midst of the seventh great extinction, a rapid, planetary die-off of millions of species, this time caused by humanity. The oceans are being overfished, the forests shaved, and wildlife exterminated. Many species of flora and fauna that sustain natural systems and, as a direct result, human life are vanishing before our eyes. Although I care about protecting wildlife and find it personally meaningful to be in places that have not been degraded by man, today the real battle for wilderness is the battle for functioning ecosystems and the stability they provide. Worldwide environmental destruction and resource depletion mean a grim future or no future at all for billions of lives, human and other. In the words of Jane Goodall, "We've just been stealing, stealing, stealing from our children and it's shocking."

For example, Lester Brown, in his book *Plan B 3.0*, observes that "Haiti, a country of 9.6 million people, was once largely covered with forests, but growing firewood demand and land clearing for farming have left forests standing on scarcely 4 percent of its land. First the trees go, and then the soil. Once a tropical paradise, Haiti is a case study of a country caught in an ecological/economic downward spiral from which it has not been able to escape. It is a failed state, a country sustained by international life-support systems of food aid and economic assistance."

The same story is unfolding on a global scale, a case in point for Trent's theory of the commons, and how we are destroying what sustains us and exterminating the other creatures we share our planet with. The Native American adage that "we do not inherit the earth from our ancestors, we borrow it from our children" is more relevant today than ever. Alternatively, if you prefer the hardened scientist's perspective, Carl Sagan put it in its simplest terms when he wrote, "Anything else you're interested in is not going to happen if you can't breathe the air and drink the water."

Despite the scope of this environmental crisis, mainstream news sources around the world virtually ignore developments related to the species and systems that sustain us. Day to day we bask in the shameless narcissism of our own species, viewing and reading superficial accounts of sports, politics, celebrities, economics, wars. Even in the context of religion the view is remarkably anthropocentric. Our religions seem devoid of any real focus on protecting what gives us life. In Christianity, the Ten Commandments direct various elements of human-to-human interaction but say nothing of stewardship and respect for nature and nonhuman beings. Across the board, in all the major religions, it seems that God skipped that topic.

I can remember when my only concept of the destruction of nature came through books and documentaries, or places like the Bronx Zoo. Often I would stay awake late in my room at night reading the work of the conservationists I admired. These were dispatches from people who had seen firsthand the world being picked apart, who had worked with imperiled species, seen the logging roads, the mines, the mountaintops blow. I had felt their urgency, and so had begun my own quest. I had left the eastern United States, where not a single scrap of original-growth forest exists amid the human landscape, and saw a similar east in Brazil's Atlantic coast. In India I had been lucky to walk in the scant remaining forests and savannahs, mountains and backwaters, fantastic places where wildlife and humans were competing for space in an ever-transitioning world. And so I had seen with my own eyes the scope of destruction and what inevitable future lay ahead. I had seen the mountains removed by mines, rivers choked by garbage and razed by dams, forests cleared. Then, of course, there was the Madre de Dios, where so much of the land seemed to remain just as it had for millennia. There I had taken refuge in the sprawling majesty of the jungle. Yet the sum of this travel and study, and the ground-level perspective of the biosphere it had afforded, painted a bleak picture of things to come.

They say that ignorance is bliss, and it must be true, because the things I saw haunted me. There had been a time when the Madre de Dios, never mind the Amazon as a whole, had seemed too infinite to fall by the toiling of ants and men. There I had felt safe. Somewhere in my mind was the mistaken belief that if it had survived so long, it would continue to do so. I was not the first to make such a naive assumption.

After decades in Africa, author and artist Peter Beard wrote: "I could never have guessed what was going to happen. Kenya's population was roughly five million, with about 100 tribes scattered throughout the endless "wild—deer—ness." It was authentic, unspoiled, teeming with big game—so enormous it appeared inexhaustible. Everyone agreed it was too big to be destroyed. Now Kenya's population of over 30 million drains the country's limited and diminishing resources at an amazing rate: surrounding, isolating, and relentlessly pressuring the last pockets of wildlife in denatured Africa. The beautiful play period has come to an end. Millions of years of evolutionary processes have been destroyed in the blink of an eye. The Pleistocene is paved over, cannibalism is swallowed up by commercialism, arrows become AK-47s, colonialism is replaced by the power, the prestige, and the corruption of the international aid industry. This is The End Of The Game."

Beard's words were written long before the 2010s, when the elephant slaughter really picked up, when 60 percent of Africa's forest elephants were exterminated. Before China's legalization of elephant ivory created international demand that sent poachers wielding rocket launchers to shoot down the pachyderms. Before they began using axes to chop off the faces of mothers to get at the tusks as they screamed, their young beside them. Before a highway through the Serengeti was an idea. Before humans ate Africa.

What is it about our species that allows us to watch sitcoms and argue over sports while cultures and creatures and those things meek and green and good are chopped, shot, and burned from the world for a buck? Why is it that we so revere compassionate, heroic characters in literature and theory but practice such arrant apathy?

When the highway and road came resurrected to the Madre de Dios, I feared the worst for Amazonia, but my first reaction was to rise up and fight. But Emma had chided me for being overly naive and American in my view of the world; she said the road was inevitable and there was nothing we could do. "Paul, this isn't the movies, mate. Sometimes the bad guys win."

The insidious inevitability of progress, the cancer of the planet, had reached western Amazonia. It seemed that everything had erupted in an instant, and there was no hope. I was left feeling gutted. Was the highway the beginning of the End of the Game for the Amazon? Why couldn't I have been born a century earlier in history, when such large-scale holocaust was unimaginable? Could it be that my life's work would be nothing more than documenting the destruction of nature's greatest creation? With black plumes bleeding into the sky on the horizon and the groan of chain saws, there seemed no denying it—the west was being won.

If there is any way to accurately convey the isolating anguish pulsing through my mind and heart during this time of bleak news, it lies in an event that took place on the fringe of the North Atlantic, on what seemed to be an ordinary winter night in New Jersey. I was in my last semester at college and was just preparing to hit the sack for the night when the ten o'clock news reported that there was a beached humpback whale calf in Southampton, New York, far out on Long Island. It was the first time a whale had been beached so close to home, and within forty minutes I was flying up the highway on the three-hour drive toward the whale. For some reason, I had to see it.

It was almost 3 A.M. when I stepped out of my car and was bludgeoned by the frigid, chaotic wind of the wintry beach. As I walked through the inky blackness, salt air filled my nostrils. I had made the entire journey, skeptical of actually finding the whale, but it was not difficult to find the yellow tape that restricted a large area of beach. As I ducked under the tape I heard it for the first time, a thunderous baritone moan that resonated in my ribs. I felt my blood go cold with sudden apprehension, as the reality of seeing a doomed leviathan began to sink in. For a moment, I thought of turning back, though my legs continued forward. A sudden blast rang out above the din, the disembodied shock of a blowhole exhaling. The caliber of it, the tremendous quantity of air being moved, ignited a very primal fear that sent my knees shaking as I cautiously approached.

Crossing over the sand to where the waves broke, I saw the whale rocking in the surf. It was a sub-adult, not more than thirty feet long. His long white fins worked to steady his large black torpedo body, as waves repeatedly crashed and sprayed over him. I approached with the cautious steps of one not sure whether he is dreaming or awake. As I stood just feet from the whale, its large wrinkled eyelids framed a single huge retina that stared sadly into my own. Again the whale groaned, and shook the earth, the air, and my ribs.

Rocked by a wave the whale pitched and violently slapped its tail to the ground for balance, and then gasped. It was a staggering display of power. But when I advanced the whale was still. We stared at each other as I drew nearer, cautiously stepping into the frigid surf. To get near enough to touch him I had to wade up to my belt in the winter ocean amid the sting of the spray. Now beside and beneath the rocking giant, I placed both hands on its body, a smooth rubber enormity. For a moment I felt the whale and the whale felt me, there on the border of the dark infinity of ocean.

Driving home in silence on the deserted highway, my heart was broken for the whale, the groans of agony of its labored breathing still playing in my mind. In the end, it would take a week for the whale to die, suffocating slowly under its own weight, all alone. Although I was not able to help the whale, something in me was deeply grateful for having been by its side on the worst day of its life. Whether the cetacean registered my compassion in the moments that we touched is debatable, but what I know for sure is that it means something that someone cared. If something so great must pass from existence and no other solace exists than bearing witness, then so be it.

For years I had insulated myself in the Madre de Dios, almost greedily coveting my own Eden on the Las Piedras. I realize now that what drove me there was the search for a place where the world had not been destroyed by the sprawling, teeming greed of my own species. There I had found adventure and pure beauty, my little anteater; I had played in paradise and for a moment even thought it might last. The road changed all that. From the time of the great basin's inception to the present, over the hundreds of millions of years that make up the Amazon's creation, there is no doubt that we are alive for the century that will decide its fate; the Battle of the Amazon is on.

I visited Don Santiago on my own and asked him to draw me a map of the route to reach the Western Gate. He was hesitant. "At this time of year, with the rain, it's impossible," he warned. Still I begged him to show me, and he obliged, drawing crude lines into my notebook with his gnarled fingers and providing an afternoon's worth of oral history on the journey to get there. "But you have to be careful," he warned. "The jungle at this time of year is dangerous; you don't understand its power." I told him I understood, but was inadvertently lying to both of us.

As if in a dream I packed a bag of bare essentials: machete, hammock-tent, pot, lighters, compass, and then spent two days traveling as deep into the forest as I could get with humans. I was able to hitch a ride up an incredibly remote tributary of the Tambopata. Disembarking the boat beneath the towering canopy, I waved thank-you to the men who brought me and made some minor adjustments to my pack.

Had I been thinking rationally I would have noticed the thick cloud ceiling that had gathered angrily above the canopy, and might have paused for a moment to think about the gravity of committing myself to the mercy of the Amazon in its most violent season. Instead I turned to face the wall of dark foliage before me and drew a deep breath, then started walking.

13
Storm Solo

It's a land that God, if he exists, has created in anger.
—WERNER HERZOG, *THE BURDEN OF DREAMS*

Thunder pulsed in the burnt sky as hours and miles passed. There were no paths between the miles of mute, expectant trees. Before leaving Puerto Maldonado I had printed a map, which I checked against my compass to hold a bearing every five hundred feet. The paper in my hand showed a larger squiggle and a smaller one: the river I had left, and the one I was going to. The plan was to traverse the jungle between the two rivers in a single day, and then spend several more freely exploring. The destination river, the Tahuamara, was virtually untouched. On the map, the area between the two rivers was a seventeen-mile stretch of homogenous green. What the map didn't show, and what I didn't know, was that between these two rivers was the largest swamp system for hundreds of miles around. Even if the seventeen miles had been dry forest, the distance would have been near impossible to cover, but in the swamp maze, there was no chance. I was doomed from the start.

The first day I spent walking—endlessly. I camped at night and woke at 5 A.M. In the sullen silence of the overcast dawn, I packed up and continued on my bearing, deliberately forcing forward optimistic thoughts that I would reach my destination by noon. By 9:30 A.M. the rain was falling hard, and I was still hacking endlessly through the brush. By midday I knew I was in trouble. I should have reached the river long before. Within the swampy, tangled forest the bugs were horrendous, the worst I had ever encountered. Before me as I walked was a hovering mass of dozens of mosquitoes, their shrill wheezing drone working on my brain like a dentist drill as I fought desperately to keep the word *lost* out of my mind.

Despite my frequent compass checks, the forest had its way with my sense of direction. It was simply impossible to maintain a bearing. I would find south and start marching, but twenty minutes later when I checked my compass it told me I had been traveling northwest. It was an eerie realization to make: the compass or the brain, one was malfunctioning.* By 3:30 P.M., as it began to rain in earnest, I vaguely took note of a tremendous termite nest that had been constructed around a slanted vine, the nest hung like an upside-down teardrop. Four grueling hours later, when I should have been miles farther on my course, I came before the same nest again. At the sight of it, panic jolted through my veins.

I walked faster then, heart pounding. Claustrophobic urgency propelled every step as I craned my neck and scanned in every direction, praying for a gap in the foliage that would signal the river. The jungle was groaning and buckling, ominously growing darker as the canopy above churned in the

---

* Later on I would learn that it is common knowledge among some people that the iron and other properties in Amazonian trees can render a compass completely useless.

wind. The clatter and pound of falling debris from above were terrifying. The storm was gathering force and my heart was pounding. As the wind gusted, a hundred feet to my right a branch the size of a mature oak snapped and hit the earth with the force of a car crash. As one foot went before the other with ever increasing desperation I couldn't help but wonder what the hell I had been thinking. I needed to find shelter. If the storm opened up full blast, there was little chance of surviving the carpet bombing of shed tree limbs that would result. *Please let me find the river*, I whispered through clenched teeth as I hiked on, but as the weak light began to fade it became ever clearer that I would be spending a second night out, this time with no way of denying the reality that I was lost.

Virtually anyone familiar with Amazonian exploration has heard the name Percy Fawcett. A legendary explorer at the turn of the twentieth century, Fawcett mapped previously unknown parts of the Amazon Basin and along the way encountered tribes that had never before seen outsiders. Convinced that the jungle held a lost civilization, which he named "Z," Fawcett lived his life as a series of ever-intensifying missions that eventually cost him his life; the explorer vanished in 1925 and was never heard from again. Yet despite the remarkable adventures of Fawcett himself, the most haunting indication of the Amazon's power came in the wake of his disappearance, as expeditions were launched by teams from around the world to find the lost explorer. Expedition after expedition followed Fawcett's fate and were swallowed up by the jungle. More than a hundred people who went searching for Fawcett never returned.

As I made camp I tried to squash the ticker of thoughts deafening my mind. Measurements, deductions, predictions, superstitions, and analysis of every possible outcome. Most immediate on the list was how likely or not it was that I'd get crushed by a falling tree. Next was what would happen after tomorrow if I still couldn't find a way out. I focused on breathing and tried to concentrate on making camp flawlessly: hammock set, shoes, and pack hung from a cord to reduce the amount of ants that doubtlessly would accumulate; small brush cleared; machete on the ground beside me.

As I made camp, the image of my hammock being smashed into the earth by a falling branch while I slept played over and over in my mind. No one would ever find my body. There was nothing I could do about it, though. I needed sleep.

Inside my hammock, the break from the mosquitoes was delightful. The entire day had been filled with them, I have never seen so many in my life. Las Piedras certainly was not so bad as here, not even close. Even La Torre had been a walk in the park compared to this. Maybe it was the endless swamp, or proximity to lakes. Whatever it was, the teeming hordes of bloodsucking insects, and their ever-present drone, only increased the fear-inducing repulsiveness of the jungle labyrinth that would not release me.

For the first time in my life I felt that it was possible I might die. So many hours of persistent and fruitless walking had seemingly gotten me nowhere, and from looking at the map, it was painfully clear that the landscape was far larger and more intricate than I had ever thought. The reality was that in hundreds of miles of forest it could take weeks to find my way out. But I didn't have weeks—a few more days spent this way, and I'd be gone. The clock was ticking.

Lightning flashed green in the canopy above, seething, promising things to come. It took all of my concentration to keep my eyes shut and try to pretend that sleep was possible.

For hours I lay this way. Suddenly I awoke in pitch black. For a breathless moment I was frozen as my mind booted up, and tried to put back together where I was. It was on this night that I awoke to the jaguar just inches from my right ear—so close, I could feel her breath as she drew in my scent, her face only inches from my right ear.

Despite being entirely at the mercy of the cat beside me, I gradually began to feel a thrilling, calm wonder. In the blackness, the jaguar and the human were disembodied forms exchanging sound. Cautiously I moved my arm a half inch. Again the soft thunder from her throat washed into my ear in hot breath. For several minutes, there was no sound but that of breathing and beating hearts.

The slightest sound of flexing leaf fibers announced her departure. I moved my hand once again, but this time there was no reaction to the sound. Slowly sitting up, I unzipped my tent and switched on my headlamp. Water vapor particles hung animated in the still air in the beams of light, the trees ominous and alien above. Just moments after being inches from a jaguar I stood panting with relief and exultation in the jungle night, no longer afraid.

I traced my fingers through the soft impressions made by the jaguar's paw. Moving around camp, I brushed the ants off my hanging shoes and pack and removed some tobacco and a handmade pipe a close friend had given me weeks earlier. The smoke curled and intermingled with the vapor amid the foliage as I tried to digest what had just transpired, and what lay ahead.

When you are lost alone in the wild, when matters become serious enough that you fear for your life, it is rough having no one to share the experience with. The day had been an eternity of pounding stress over monotonous hours in horrible silence, the strange swampland seeming devoid of life. The fear and wonder of the jaguar's visit had overpowered the dread of being lost and shocked me out of loneliness. As I puffed my pipe, it was strangely comforting to think of the jaguar stalking the forest nearby, and that she had come to check on me. At least she was out here, too.

In the morning the odyssey continued much as it had the days before: the horrendous tangled foliage of the endless swamp; the angry, brooding sky; the lonely stress of being lost. I tried not to think about water. I had stopped sweating despite the humidity, which was worrisome. There *had* to be a stream somewhere. Carefully consulting the map, I changed course by several degrees, and was encouraged when around noon, the sun emerged and the land lifted away from the marshy floodplain and once again became terra firma. Progress came faster, and suddenly I was covering several miles per hour.

By 11 A.M. I had been hiking for more than five hours, as the forest again began to show signs of change. This change I recognized as good. Just over an hour later, I heard the slow rushing of water. The terra firma descended into sandy floodplain and floodplain petered out into a jungle of massive wild cane. I sprinted beneath the twenty-five-foot stalks of prehistoric-looking grasses in wild excitement before finally emerging out of the green and into the wide-open freedom of a beach.

I fell to my knees on the sand beside the river and threw my arms in the air, then fell onto my back. The clouds were breaking up and sun flooded loving warmth over my body; it felt like being reborn. It felt like I could breath again. Throwing off my pack, throwing down my machete, and peeling off my clothes, I headed for the river. I jumped and splashed and made all the noise I could. Swimming below the surface and gulping, I tried to drink the entire river.

The open beach was the antithesis of the haunted swamp I had spent the previous days in. Walking upriver along the sprawling beach, I gaped up at the blue-and-yellow macaws that flew against a magnificently clear sky. Over the course of the afternoon I saw turtles lying on riverside logs taking in the hot afternoon sun with butterflies on their noses. Monkeys leapt through the trees and capybara splashed in the river. Machete in hand, I walked up the sandy beaches, taking it all in.

Many naturalists before me have experienced this fascinating dynamic in which you come to know species so well, it creates an almost neighborly familiarity (for the naturalist, of course, not the animal). It must sound like lunacy to anyone who has not experienced it, but walking up that beach as orocerulean macaws swept across the sky in all their shrill, colorburst glory, I was glad to see them. It was the same for the capybara, turtles, and other creatures. Added to the neighbor phenomenon, or perhaps a part of it, I also was cognizant of feeling more at ease with other creatures around; if the capybara were calm, I could be calm. The same went for the turtles, lapwings, and other life on the open beaches.

It was the first time ever I was out in the jungle without any restrictions, responsibilities, or distractions. No one was waiting for me to return to the station; no one could find me even if they wanted to. I had the greatest jungle on earth all to myself; it belonged to me, and I to it. Throughout the day I drank and drank, and took frequent plunges into the river. The plunges became more cautious after I spotted a black caiman in the late afternoon that was well over ten feet.

In the glowing rays of evening I cooked out on the beach: ramen noodles, Brazil nuts, and chocolate, followed by tea. With the horizon's victory over the sun, quiet cloaked the world. Orange light faded and the sky became a quiet spectrum of blue and purple, accented by leftover wisps of orange cloud. With the elation of survival, the relief of being free, I experienced a moment I had always aspired to live out, alone in the Amazon.

That night I made camp in high hopes that tomorrow would be a peaceful day of exploration on this river that seemed ignorant of the existence of man. Despite the harrowing journey to get there, I had made it. Everything had worked out. If the weather cooperated, I planned to spend as much as a week on the deserted river, enjoying the jungle. Yet on this night as well, peace was not in the cards.

In the panic of the journey I had largely pushed the subject of the highway to the back burner, forced to forget the questions that gnawed inside me. Yet as I stared out at the world of natural serenity in the forgiving hours of that storm, my thoughts remained haunted.

At 1 A.M. the sky exploded like nothing I have ever seen. The rain fell in fat droplets and the entire world became a rushing mass. Flooding in the west had caused the river to rise more than twenty feet just in the time I had slept, and when I stepped out of my hammock I found myself in thigh-deep water. There was no choice but to untie the hammock and retreat to higher ground. It was a miserable and cold night spent sitting at the base of a tree, soaked, waiting for morning. Yet the morning changed nothing. The deluge continued with frightening power. After the days of madness in the maze, and now this, my body was beginning to feel significantly run down.

With no end to the storm in sight, I had to move to keep warm, and so started making my way downriver.

The following day the storm only worsened. Another sleepless cold night beneath abusively loud thunder had taken its toll on me. In fact, the thunder was so loud it was terrifying. Each time lightning flickered I took shelter under my arms. Booming avalanches of thunder raged in the sky just above the canopy, at such volume that my eardrums hurt. Billions of collisions between water and leaf created a ceaseless din.

The flood had swallowed all the beaches and transformed the river. Overnight it had changed from a lazy lowland tributary to an aggressive, gushing mass. As a result I was forced to hoof it through the jungle. By midday I knew I couldn't spend another night in this rain; my body was in bad shape. I was cold, soaked, and unable to maintain the calories to support the constant frenetic movement and exertion of navigating the jungle. I knew that I had gotten lucky in escaping the swamp maze; I knew how bad that could have gone. Now with the storm and my deteriorating condition, I felt hyperaware that if I didn't get out now, I might not have another chance.

As I hiked downstream the river became an impressive parade of timber. Entire trees, some as thick as a bus, floated down the racing current. JJ and I had often caught rides on logs to get places, so long as those destinations were downriver. But out here the river was fast, and the trees gigantic and very dangerous. For a while I hesitated, but as the rain continued, it seemed safer to hop on a log and get to shelter and humans than to spend another night out there in the storm. When an especially large tree came barreling by, I plunged into the water, swam to it, and climbed aboard. Such was the tree's girth that balance was not an issue, I wasn't large enough to influence its movement and could stand and even walk on its broad trunk as it was swept downstream.

Atop the massive beam I was surrounded by debris in the turbulent river. Giant timbers creaked and groaned as they collided, and every few minutes my tree would hit another giant, sending me sprawling onto my face. It was dangerous, but the travel speed was unbeatable. From the open river I watched the canopy buckling under the high winds of the storm, as lightning reached down in arcs to the land. Two hours after I climbed on, my tree had taken me back onto the larger Tambopata River, where the turbulence at least was much softer. But on the larger river, lightning was a greater concern, as was the fact that there were now thousands of huge trees swarming and smashing beside me. As I rounded a bend I could see that my tree, whose base was nearly twelve feet in diameter, was headed for disaster. From a quarter mile away I saw the logjam, where thousands of beams of every size were smashing into each other and being splintered in the current.

To jump into the river would have been suicide, so instead I jumped from log to log, making my way toward the shore. Fortunately there were enough logs that I was able to get to the river's edge without incident. The rain, at last, seemed to be lessening, though my pack and everything in it were soaked and heavy.

An hour later I had made my way to the top of a cliff that looked out over the river. I hoped someone would come by in a boat, but in that timber-choked river you would have to be insane to try that. On top of the cliff I sat in the soft drizzle, opened a bag of Brazil nuts, and began stuffing fuel into my empty stomach. For a time it seemed like the rain had at last tired itself out. However, as I ate, a drawing of wind and a rumble seemed to silence everything. The worst was about to come.

Dark clouds curled and twisted as they approached. I watched as miles of canopy were devoured in the approaching gloom. The ground began to shake. A hissing, imperceptible at first, grew in volume, roaring and lurching over the jungle as rain stampeded in a deafening siege, like the sound of a thousand freight trains. Jumping to my feet, I watched the shadow avalanche over the tops of the canopy and raze the surface of the river. The water boiled under the sky's wrath, the river's surface a chaotic sheet as the storm enveloped the world. I ran, sprinting along the tree line as the ground shook and all hell broke loose.

I was now surrounded by one of the most important biogeochemical processes on the planet. They say the Amazon is the lungs of our planet, and it is true that nearly 20 percent of the oxygen on earth is produced there. Yet in the Amazon most of the oxygen that the jungle produces is absorbed back up each night through respiration. Inhaling and exhaling, the basin is almost entirely a closed system of self-sustaining processes. It is in fact the fallen leaves that lead to the creation of oxygen in a magnificently large process.

When South America drifted away from Africa's embrace millions of years ago, their relationship was not ended. Today the two giants converse in a language as old as time, trading gifts between oceans. This takes the form of phosphorous-rich sediment dust that leaves Africa, borne on the wind. Carried over the Atlantic Ocean, the nutrient-rich particles sweep over the Amazon, where they are absorbed by raindrops and hurled to earth. Each rainy season, nourished by African minerals and watered by rain, billions of photosynthesizing organisms over thousands of miles jump into fertilized hyperdrive. As the growth explodes and leaves fall to the forest floor, their nutrients are almost entirely consumed by fungus and reabsorbed by trees (if organic matter were not broken down by fungus so efficiently, the jungle would bury itself in leaves). The torrential flooding caused by the incessant deluge turns the forest to river, where the currents sweep the jungle floor's bio-sediment of nutrient-rich detritus into tributaries that dump into the main Amazon River; as much as two million tons can be swept from the basin every twenty-four hours.

Laden with the stored energy of so many tons of organic material, the Amazon River carries its litter 4,000 miles across the continent, finally dumping it into the Atlantic Ocean. The mouth of the Amazon is more than a hundred miles wide; its bulk also pushes out nearly that far into the ocean before salt and fresh water mix. It is there that microscopic plankton in the ocean feast on the nutrients in the fresh water, absorbing carbon dioxide and releasing oxygen into the atmosphere. The plankton blooms that spew from the Amazon's mouth can cover twenty-five thousand square miles.

Even the lightning is a source of life. Blasting through the air at five times the temperature of the surface of the sun, lightning rips apart air molecules that reorganize into nitrate, which join the phosphates from African dust that dissolves into raindrops and falls into the jungle below.

Lost and desperate in the apocalyptic storm, I was at the center of the engine, a violent, cataclysmic act of creation. The tons of leaves and trees speeding past would continue across the continent into larger and larger rivers, traveling down the body of the Amazon to be released from the great river's mouth, like a giant serpent breathing life into the earth.

As the rain overwhelmed the earth, the jungle purged its

contents into the river. Along the riverbank I ran, screaming at the storm to give me its worst, in adrenaline-induced
madness. Hurricane winds ripped trees from their roots, and
splitting timber echoed the thunder as giants were thrown to
the ground. As the jungle tore at itself, lightning added to the
genesis, lacerating the sky, producing ozone and nitrate-rich
raindrops that pelted the jungle below. The vital nutrients produced by lightning each day across our planet are absorbed by
vegetation and consumed by animals, including humans; that
night, wicked purple fingers ripped across the sky, slashing at
the treetops, overwhelming the night in overexposed snapshots
of the storm that were burnt into my eyes for moments after.

The thunder was so bad, so monstrously close overhead,
that at times I fell to the ground. Like cosmic Velcro tearing
in the heavens, the thunder roared, cascading. *Flasssssssh*-
BOOM! It felt like the world was ending. In the surreal dimness beneath the clouds, raindrops turned visibility to almost
nothing. The storm raged on and I ran. There was no way to
stop; I couldn't. Looking back, I think it's easy to claim that I
ran for those hours amid the madness because it would keep
me warm, because if I stopped I'd be shivering and screwed
within moments; but in reality I didn't know why I was running, and I still don't.

In *The Call of the Wild*, Jack London wrote, "There is an
ecstasy that marks the summit of life, and beyond which life
cannot rise. And such is the paradox of living, this ecstasy
comes when one is most alive, and it comes as a complete for-
getfulness that one is alive." The pounding reality of the storm
demanded the full attention of my every faculty and I slipped
into that magical plane where thought and consciousness fade
into present action.

Though in hindsight the experience seems somewhat romantic, in reality I was in trouble. Moving through the jungle was deadly dangerous as limbs fell, and the muddy banks of the river frequently sucked my body up to my waist. Finally, while I was attempting to traverse a sixty-foot cliff by holding on to a root, the bank gave and I went tumbling down. Buried so badly that for a moment I thought I would suffocate, I had to fight to the surface, where I lay panting. It was over.

The storm continued as I lay prostrate. My machete was gone. On hands and knees I crawled over the mud, too broken to stand. Beneath a tree, I drew my knees up to my chest as my body began to shiver violently. I knew I was in trouble. My watch said it was eight o'clock at night, which meant it was only going to get colder. Nine hours until morning.

With no shelter I shivered in isolation on the bank of the river. My body was covered in deep gashes, cuts, and bruises from the mudslide. James Murray, alumnus of the legendary Shackleton Antarctica voyage of 1914–16, once said of the Amazon, "Without a machete, it means death to be lost in such a forest," and he was right. Without a blade there was no way for me to make shelter. Without some way of getting out of the rain and getting warm, I was in very grave risk for exposure. How I cursed myself.

One hour passed, then another as I lay in the mud. I was shivering uncontrollably. I tried to think of some way of . . . something, anything to get out of the rain. But I was too weak, and there was nothing but jungle. If I had made it farther downriver on that log, it would have been possible to find help—the lower Tambopata has plenty of farms. The truth was, though, that I had no idea where I was.

I had always been lucky. Every time it had seemed like the hammer was going to fall, I had been saved. The jaguar just two nights ago. Hell, Gowri and everything that happened in India in the twenty-fifth hour, JJ showing up just as my ribs were about to be collapsed by the anaconda on La Torre . . . But, lying in the mud, I realized that this time I wasn't getting bailed out. There was no one *to* bail me out; it was only me. For the first time, I was in real trouble with no chance of being saved; it was going to be a very long night.

I lay in the rain, slipping in and out of consciousness. My shivering woke me at 11 P.M. I had only lain in the mud for three hours. The rain had slowed but the cold had worsened. Wrapped in my hammock, I shivered convulsively. My teeth chattered and I hung my jaw loose; I groaned through the clacking. Trying to sleep was miserable, but I was determined to keep my eyes shut. Time wore on.

When I first saw the light I thought I was cracking up. One second it was there, then it was gone. For a long moment, I stared into the blackness. Then it flashed again, and my pulse hastened: *humans, shelter, warmth!* It was more than a half mile away, too far to yell. I stood up and grabbed my headlamp and set it to the rescue setting: flashing on and off. For five minutes I stood with my hand high above my head, hoping the light would see me, but with each passing moment it became clear that they were simply too far off.

In despair I let my hand fall. Whoever it was, they were not looking my way. *Please let them look in my direction just once!* It looked like a powerful spotlight, which was confusing. Locals wouldn't normally have such a powerful beam. Could it be that JJ had rallied people to come rescue me? Even from a half mile away, I could see that they were using the light to scan the banks of the river. But that also didn't make sense. I had told him I'd be gone for at least a week.

The river was swelling and where I had once been standing on solid earth, I was now in knee-deep water. Ghostly black trees floated past me with menacing velocity. Then suddenly light swung in my direction. I desperately waved and shouted with my headlamp aloft. The mysterious light was pointing directly at me for some moments, before it extinguished completely: then I knew that they were looking at me.

For the next forty minutes I kept waving and screaming as the mysterious boat continued to scan the jungle and riverside, occasionally checking in my direction. It was more than an hour before they grew sufficiently curious and came toward me. Blinded in the light of their torch, and covered in blood and mud, bent and filthy, I waved my hand to whoever might be on board. I was shielding my eyes from the torch beam when a voice came from the other side of the light: "Hola." It was clear that the boat was having trouble maneuvering the rapid current and the timber. "Hola!" I called back over the din.

A skinny young man shouted to me from the boat, "Where are your friends, em . . . your guide?" I smiled and told him I was alone. He assumed I didn't understand the question and repeatedly asked where my guide was. But after getting nowhere, finally he asked if I would like a ride. Catching a rope and pulling the boat in, I grabbed what was left of my gear and climbed on board.

I stepped into the boat and limped past a couple of German tourists. "Hey!" I said with a friendly grin, but they both averted their gaze and said nothing. They didn't seem at all amused; instead they appeared almost frightened of me (to be fair, I was covered in mud and blood, and my clothes were rags). We shoved off and I learned that the wiry guy was a guide. On board were also two German tourists and a boat driver. They had been out scanning for caiman in the storm because it was the Germans' last night in the jungle with Inotawa Expeditions, before heading out to Machu Picchu. Once I had been seated on a gasoline canister, the guide politely asked if I was fit to sit tight while they finished their hunt. I told them that was fine.

I sat shivering but happy as we drove; I was just glad to be out of the rain. The guide finally caught a baby spectacled caiman and presented it to the tourists. He asked the wife if she would like to hold it, but she declined and instead took a photo of her husband holding it up. What happened next was a moment of awkwardness in which the guide did not know exactly what to do. He turned to me and asked, "Would you like to hold it?"

"No, gracias," I replied, smiling in the dark. Then the German woman turned to me. "It's okay," she said. "I was scared to hold it, too!"

We traveled downriver and that night I was given a castaway's welcome at Inotawa Lodge. My clothing, camera, and other ruined possessions were dumped, and the staff provided a hot shower and loaned their own shirts and pants for me to wear. They brought me tea and invited me into the kitchen while the many tourists staying at the lodge ate in the dining area. I watched in dazed disbelief as the staff placed warm soup, coffee, tea, bread, cake, cigarettes, and chocolate before me. And as I dug in, they asked questions and tourists snuck glimpses through the door. Again and again I thanked them for their incredible care and kindness. When I had finished enjoying the new faces and many gifts, I excused myself and limped slowly to my room, where a deliciously clean, safe, dry bed awaited. There I fell into a deep and carefree sleep for the first time in a week.

14
Poaching Poachers

Tightly choked pump-action scatterguns with heavy pellets were our weapons of choice against poachers, for in the dark, in the bush, things are about as close and personal as you can get.

—LAWRENCE ANTHONY

Two days later, when the storm finally broke, an Inotawa boat bound for Puerto gave me a lift. I collapsed outside JJ's house and waited. He arrived with eyes wide *"Pauool!* Oh dear, I thought . . ."* He hugged me. "It was so bad! I was thinking, and seeing the *terrible* storm and thought oh, he is in biiiiiig trouble." Pico came limping into the yard a moment later with a similar reaction: *"Puta Madre!* You look like shit!" We moved to a café and as lunch was served, I told them what had transpired, and how Inotawa had rescued me. They took in the story with relish.

When I had finished, however, I noticed that both brothers were staring at each other and they began conversing rapidly. Though I couldn't follow completely, I gathered that there was something that wasn't making sense to them. Finally, JJ, looking confused, asked, "How did you get on the Tambopata?" I explained my route and showed him what was left of the map. Pico threw his head back laughing as JJ's shoulders began to shake. They hooted, howled, and explained that it had all been for nothing. Apparently I had assumed that Don Santiago's sketch meant one river, when in fact it meant another more than forty miles away. "He wasn't even close!" Pico spat and coughed in laughter. I began to laugh too, though my face and everything else hurt.

I had fled into the wild in search of answers and had been hammered and broken down by the jungle and storm, brought to the edge and then back again, and emerged reforged. I had experienced Amazon in brutal, intimate communion, and lived. I was acutely aware of being *alive*. Everything felt illuminated. Perhaps it was residual adrenaline or profound relief, but for two quiet days I had drifted in and out of consciousness at Inotawa, contemplating what I had just experienced, and the world seemed brighter. My fear and frustration had been replaced by a powerful and almost irrational hope.

In those quiet, rainy days at Inotowa, I had patiently burned to see my family, Gowri, and to live. I felt like I had unlocked some secret of consciousness, and felt as if anything was possible. Hours were spent scribbling realizations and plans into my journal. The positive charge I felt in the wake of that solo expedition left an afterglow that lasted for years. Despite having almost died, and the fact that I had not been anywhere near where I was trying to go, the solo was an adventure, a recharge, and I emerged determined to do many things— determined to live.

Over the next year I kept my promise to Gowri and returned to India. Together again, we continued what we had started months earlier, engulfed in the mysterious magnetism between us—which the months and thousands of miles between us had failed to break.

After we were reunited, our adventures only escalated and took on an almost cinematic grandeur—climbing the orange boulders of Karnataka, exploring the green jungles in Kerala, nights by the Bay of Bengal. Visions of Pondicherry are forever painted in my mind—running in the evening with Gowri through the bustling cobblestone streets of vendors and traffic in the salt-sweet coastal air, alongside a temple elephant. We held hands as we dodged the traffic, and marveled at the behemoth, its trunk swinging from side to side as it went, bells jingling from its ankles. However, for me the elephant and the vibrant insanity of the street were merely a frame for her. In flame-orange Patialas and forest-green curta, black hair swirling about her smiling face and dark eyes—it was almost more than I could take.

There was no longer any doubt that the universe had made the inexplicable decision to unite two parts of a whole from opposite sides of the world, to create something written by fate. We got engaged in a tent, beneath a powerful thunderstorm, amid the dark mist-shrouded incisors of the Agumbe rainforest of the Western Ghats. We formed a plan for her to come for college in the States and began the long process of making it a reality, first with her parents, then with mine, and then with colleges and U.S. immigration officials.

While this was going on, JJ and I set our sights on a new project: developing Don Santiago's thousand-acre property into a rainforest reserve and ecotourism destination. We were starting anew. Though secretly my heart burned for Las Piedras, and nowhere else, I dived into the Infierno project with everything I had.

I began reading the work of great conservationists who had succeeded in protecting large tracts of habitat all over the world: Alan Rabinowitz, Charles Munn, Gabriella Fredriksson, George Schaller, Lawrence Anthony, Paul Watson, Jane Goodall, Steve Irwin, and so many more. These were people who had employed ingenious methods of raising awareness, leveraging governments, and mobilizing locals, employing whatever tactics were possible to create ecosystem-level change to safeguard the earth for humans and animals. Through them I began building an arsenal of practical conservation knowledge, armed as well with the hope that if they had done it, I could do it.

While I was studying larger-scale conservation, I began joining Pico, JJ, Mario, José, and the entire family for hours and hours of meetings and talks, days of exploration and planning, and before long we were building a new research station on Don Santiago's riverside land.

With Emma and Joseph permanently back in the United Kingdom, JJ was gravely affected by their absence and was not the same friend I had once looked up to. There was anger in him, a distance, and it simply wasn't the same between us. Yet he did his best as we worked together to get our new project off the ground, and bridged the gap by working harder than ever. The first order of business was cleaning up the country: Santiago's large land was right next to Infierno and each night we heard numerous gunshots as hunters took what they pleased from it. In meetings with the entire Durand family, we came to the conclusion that if there was any hope of running a quality ecotourism operation, we'd have to put a stop to the poaching immediately.

So JJ and I began poaching poachers: camping out in the jungle, walking the trails at night, and tracking gunshots and footprints in the darkness. We sought the assistance of local police, but they were unwilling to send officers traipsing through the jungle and instead advised us to buy a pair of handcuffs and bring in whomever we bagged. It was rough and dangerous work, and we had more than a few close calls.

One time I found a poacher walking with a peccary over his back and played the lost tourist trick, asking him for directions and in the course of conversation, finding out his name. That was a lead, and now we knew who to watch for. JJ and I made a good team and would track hunters from opposing angles, surprising them in the act. We had each other's backs.

We worked hard to track the poachers on Santiago's land. After all, if they kept killing the wildlife, we would have little chance of starting up a conservation-ecotourism project there. We continued spending long nights in the jungle, and eventually our efforts paid off. One morning while we were doing transects with a group of volunteers, taking careful data on the wildlife we observed, someone heard the moan of chainsaws in the distance. We followed the sound and saw a group of men who had recently felled a tree. JJ instantly recognized them as belonging to the same family as the guy I had encountered with the dead peccary. The poachers were not thrilled about being caught, and if JJ and I had come across them alone, it almost certainly would have turned into a fight. But with volunteers present the men were too stunned to retaliate. With cameras rolling, we kept it calm, and later showed the footage to the police back in Puerto Maldonado. The police followed up by investigating the downed trees and poached lumber, and eventually contacted the men responsible and issued a summons and a warning that any further offenses would be punished with jail time. It was a huge success for our new project, and it sent a message to the rest of the community that Santiago's land was now a guarded wildlife sanctuary.

In the two years that followed, many events transpired, none as significant as the day Don Santiago boarded a boat with his grandchildren and sons to travel to his house after a family party. Sitting on the boat, traveling up the Tambopata, as he had done so many times throughout his life, he closed his eyes as the sun fell on his face and drew one last, peaceful breath before slipping away. His death shocked the community and the family. At his wake, it seemed like all of Puerto Maldonado was in attendance. I spent many days at Doña Carmen's helping with food, looking at photos, and sharing memories. One rainy day I sat with Pico for hours as he tossed back beers and told stories of his father, whom he loved so much. I held him as he wept through the day.

With Santiago's passing, the Infierno project took on a deep sense of purpose within the family, and as beams were erected and nails hammered, I began preparing to bring a group to the new location by recruiting up north. We continued to struggle with the poaching, and with Santiago gone, the brothers had no central leadership, which was a source of mounting tension. Regardless, we forged ahead.

Back at home I was nearing the end of college. By then both my family and Gowri's had accepted that this transplanetary relationship would somehow have to work. She had applied for college in the United States and then gone to the visa office in India, only to get rejected not once but twice. In a relationship defined by long absence, difficulty, and much worry, things reached a head, as it seemed she would never get her visa. Her parents, who had been tolerant of our cause, were losing faith and patience, and everyone said that once you fail an interview, let alone two, the chances of getting a visa drop to nearly zero. For a time, it seemed like I would have to choose between the Amazon and the perfect girl.

Rudyard Kipling wrote that God could not be everywhere and so he created mothers; mine swept in with her unique tenacity for results and began writing letters to the State Department, college professors, elected officials, and anyone else whom she could get a bead on. With hope in this new strategy, and another trip to India for me, we put it all on the line and went for one last interview—this time she passed. It was an overwhelming success after an epic struggle. Bravely leaving her family to live across the globe with people she'd never met, Gowri left India for the first time and arrived in New York City in December 2009. She met my family and friends in a big welcome celebration, in a world covered in something she had never seen before: snow.

With a major hurdle finally behind us, I began earnestly searching for answers to how I could create real change in the Madre de Dios. I began sending out feelers, making contacts, and exploring grant agencies and larger nongovernmental conservation organizations. One of the many queries I made was to the website Mongabay.com. For years I had read Mongabay, without question the best source for environmental, wildlife, and rainforest news on the Web. They responded with interest to my experiences in the west Amazon and with the floating forest, poachers, and ecotourism; a month later they published a full-length article and interview featuring me in their Young Scientists series—an honor whose true value I would not grasp until later on.

In Peru, Gowri joined me, JJ, Pico, and the rest of the family in working on the Infierno project. She took to the jungle like a fish to water, so much so that her knowledge of birds and butterflies soon surpassed even my own. For years I had dreamed of sharing the Amazon with her and seeing her wide-eyed reaction to the endless immensity of the jungle. Yet there was one place in particular that she needed to see: Las Piedras. We contacted the organization that owned it and arranged to visit for several days. It was a living dream to bring her there. As nice as Infierno was, for me, Las Piedras would always be home.

We spent a magical week exploring the trails and aguajales, the ancient trees and hallowed meandering streams—watching macaws and basking in the warm reality of being together. Yet there were a number of disturbing changes that had occurred in my absence. The organization that had bought Las Piedras Station clearly had no idea what they were doing and had let the place languish. Their incompetence was so great that even longtime clients, big-money annual companies that had been coming to Las Piedras for years through Emma and JJ, broke off ties with the place. The trails had overgrown, and, worst of all, the old man they had hired to watch the place when no one else was around had shot at every peccary and spider monkey he saw.

At one time you could barely step outside without encountering herds of boisterous boar nearly a hundred strong, but now the forest was silent. The herds had left, as had the troops of spider monkeys. A single hunter had changed the forest, profoundly altering the ecosystem. As I lay with Gowri in my arms, in the same hammock where I had once rocked Lulu to sleep, I looked about the most beautiful place I had ever known and dreamed of one day, somehow, winning it back.

As for the road downriver, where the forest had been burned and cleared two years back, there were now new houses popping up each day. With each hut came another few hectares of jungle cleared. It was straight from the rainforest destruction textbook, chapter: Roads. Yet I swallowed the bad and pushed forward, making positive progress where I could. In the spring of 2010, the Durand family and Gowri and I welcomed our first full crew of volunteers to the Infierno station. We called the operation Saona Expeditions, using the Ese-Eja word for anaconda.

The name was partly due to the fact that on an early assessment of the land, JJ, Gowri, and I were stalking through the forest, taking inventory of everything we encountered, and had an incredibly rare run-in with an anaconda. We had marveled at the massive kapok tree that Santiago had preserved, we had seen Spix's guans, Amazon coatis, and were following a herd of collared peccary when we came to a colpa, bordered by a small stream. There we pondered how the herd of pigs could have evaded us so completely, and inspected the colpa. It seemed active with many animals: tapir, peccary, deer, and I noticed a stunning blue morpho lapping at the salts below.

Climbing down to get a photo, I was beside the stream when I heard a muffled pop. Just three feet to my left, in the stream, a female anaconda more than fourteen feet long lay coiled around the body of a peccary. The herd had come through here, and the anaconda had grabbed one by the cheek as it passed and wrestled it into the water. The pop I heard was the pig's spine breaking in half. Gowri almost lost her mind at the sight of such a massive snake, but was still able to snap a stunning photo of the scene.

When JJ and his brother Federico arrived the commotion

was too great and the huge snake bolted, leaving the peccary in the stream. It was an unfortunate accident that really was my fault. I had not realized the snake was beside me until it was too late. In the end we took the freshly killed peccary back to the community, where a feast was served, compliments of the anaconda. If it weren't for Gowri's photo, no one would have believed us, and with it the atmosphere became one of rare excitement.

During May and June our first groups arrived with people from all over the world: Belgium, New York, South Africa, and various other points on the map. JJ and I were the guides, Pico the driver, Elías and his wife, Elsa (the one whose father had been eaten by the anaconda), were the groundskeeper and cook, respectively, with support from the rest of the family. The volunteers saw tamanduas, caught caiman, swam in the Tambopata, and began building transect data on the new forest block. The operation seemed a huge success, and the Durands were overwhelmingly pleased.

After weeks of preparation and a solid month of running groups, I returned to Puerto with staff and volunteers, and as usual made a call home. My dad's voice came on the line and he asked me if I was sitting down. Right away I feared someone had died, but then noticed the smile coming through in his voice: a television network that filmed for all the major networks had called and wanted me to call them back as soon as possible. Thanks to the article on Mongabay, the adventures JJ and I had had at the floating forest, with Lulu, and the struggles we'd faced in conservation had made their way around the world. I'd received mail from people from all over the globe who were fascinated and curious to learn more about the mysterious floating forest and other secrets of Amazonia. Now it was a television company from the United Kingdom; they had read about my work from an interview, and wanted to know if I'd be interested bringing viewers into the jungle through networks like Animal Planet, Discovery, and National Geographic.

While so many exciting new things were taking place, I was at heart still concentrating on the bigger picture in the Madre de Dios; the highway was still coming. Of all the conservation literature I had read, the story that remained fixed in my mind was still the creation of Bahuaja-Sonene National Park, and the astounding brilliance of the La Torre that I had experienced as a result. Charles Munn had imagined the park and provided the scientific muscle for the plan, but in the end it had been the local support and then Daniel Winitzky's documentary that had really won the battle. He had done something that the scientists, locals, and every other conservationist were unable to do: he had gotten people excited.

The documentary *Candamo*, which resulted from Munn and Winitzky's collaboration, was well received in Peru and even overseas. Reporters flocked to catch a glimpse of the living Eden in the Amazon, and suddenly the battle over the Candamo valley drew international attention. As time went on, it gradually came into focus that Las Piedras, La Torre, the Tambopata, all these incredible places that were allowing me to live out my own dreams existed because people before me had protected them.

The Candamo campaign began merging with my dream of journeying to Santiago's Eden, past the Western Gate. For years my grand solo had been placed on the back burner, while the project with Infierno and drama of getting Gowri took precedence. Yet the plan still burned within me, and as time went on, I became increasingly fascinated with the idea of filming the journey.

If what Santiago had said was true—and I knew he was never wrong—then the landscape up past the Gate would be incredible. If I could capture even a small glimpse of that world, perhaps it would be possible to follow in Winitzky's footsteps; perhaps I could get people to realize what was at stake. Given the television networks' new interest in the west Amazon, it seemed worthy of serious consideration.

Several issues immediately surfaced, the most obvious being whether camera equipment would somehow detract from the experience of the journey. This, however, was a worry I quickly discarded. Filming would provide a perfect excuse to spend days in hides, carefully observing wildlife, and the potential for positive change that could come from a documentary of this type was well worth whatever cost. Questions of cameras and conservation and surviving in the wild serendipitously led me to inevitably ask, as I had so many times before, "W.W.S.D.?" What would Steve do?

When I was in middle school and high school, I would come home every day to watch Steve Irwin's adventures. World famous for his work as the Crocodile Hunter on television, Irwin was a dedicated conservationist and first-class outdoorsman who used his fame as a platform to protect species and ecosystems, and to inspire others to do the same.

What most people don't know about the Crocodile Hunter is that before he was famous, Steve Irwin spent years working for the Australian government, rescuing and relocating crocodiles that otherwise would have been killed. In a boat with his dog, Irwin patrolled the backwaters trapping and wrestling the largest reptilian carnivore on our planet, the saltwater crocodile. The guy was the real deal. Living solo in the bush doing the most dangerous job imaginable, Irwin would often film himself working, a habit that later on would help catapult him into the international spotlight as an ambassador for the natural world.

It is impossible to know what forces allow one life to "reach across time not merely to inspire but to mold the dreams of another," as Wade Davis writes in *One River*, or to fathom the connection that can exist between two people who never met. Steve was a teacher throughout my formative years, one of the greatest guiding forces in the process of wringing fruition from dreams. I watched as he expertly handled snakes and other wildlife and would later recall his skills as I developed my own over the years. His overwhelming passion and enthusiasm for protecting life on earth had even inspired me to apply for a job at his zoo, which I had been pursuing for months, up until the time of his death. I was determined to meet the man. That I never did will always feel like an injustice.

As the solo approached, I dived into Irwin's work, concentrating on the passion he was able to transmit, even more than his skill with wildlife. In face of the cynicism and fatalism that are all too common, he was a stubborn source of light. He conquered the world with inimitable raw optimism and passion. By caring so freely, he enabled others to care, and to stand in defense of all that is green and good in the world. His influence created many conservationists, including me.

With Gowri and my friend Norm, I began filming wildlife and practicing being before a camera. From early tests with video, I knew I could never match Irwin's presence on film—no one could—but there were other ways of incorporating hope and energy into conservation. Yet no matter how deep my fascination and ambition for the Western Gate, the greatest question remained logistics.

After the last solo attempt, I had respect for how utterly helpless a single human being is in the Amazon. In the intervening time I had continued to hone my skills in the bush, accomplishing minor solos without incident. This gave me confidence that the storm solo had been a unique and isolated event, and if I went searching for the Western Gate, logically, especially if it was not the rainy season, then it should be no different than any hike I took around the station (except that it would last for days and possibly weeks).

The problem with finding the Western Gate was that the route required traveling into country so remote that every danger was amplified. I'd be traveling up a medium-sized tributary, then traversing jungle, hiking up another tributary, and crossing the Western Gate into the unknown. As I had spent two years poring over the details, it seemed doable. Then again, climbing Everest is doable—it's getting down that is often the crux. I continued to struggle with the hard fact that if something went wrong out there, it would all be over. Even something as minor as a broken leg, an infection—anything. Once I was dropped so far from civilization it could take weeks if not months to navigate back on foot after a catastrophe. How could I put myself in such a position?

The breakthrough came one day while I was reading an issue of *National Geographic*, in an office building in Manhattan, of all places. The article was about the adventures of long-distance solo trekker Andrew Skurka. The article mentioned that Skurka had hiked more than four thousand miles across eight national parks in Alaska, covering several hundred miles with a raft he carried in his pack. My heart instantly began racing and my hand gripped the peccary tooth that lay across my chest. Later that day I began researching the rafts and found a small company out of Colorado called Alpacka Raft that specialized in making white-water-capable, super-lightweight *pack-rafts*. I had just found a tool that was designed for the specific purpose of accessing inaccessible places on rugged expeditions.

It was a discovery that required a complete reevaluation of my dream expedition.

With the pack-raft I could hike as far as a river would go, deep into headwaters unreachable to anyone else, and when the time came to turn back, the raft would sweep me downstream at quadruple the speed of walking.

With a pack-raft now part of the plan, the expedition I had always dreamed of began to materialize. While leading trips with JJ, I made an effort to learn every edible and medicinal plant he could show me, a crucial backup to the three weeks of food I could comfortably carry in my pack. Well practiced with camera gear, rations calculated, pack-raft purchased, I was as ready as I ever could be to take the plunge.

15
The Launch

Deep in the forest a call was sounding, and as often as he heard this call, mysteriously thrilling and luring, he felt compelled to turn his back upon the fire and the beaten earth around it, and to plunge into the forest, and on and on, he knew not where or why; nor did he wonder where or why, the call sounding imperiously, deep in the forest.

—JACK LONDON, *THE CALL OF THE WILD*

Morning mist hung over the jungle as worlds passed by. Into the mouths of ever-narrowing rivers, toward the deep unseen, the small craft traveled. Farms and villages became fewer by the day, winnowing to the last desolate outpost towns with hopeful names, the fringe of human presence on the frontier. A woman washing clothes as her children frolicked in the warm current, a man paddling a dugout canoe piled high with the wealth of the land. In the mist a young girl with fish-shaped eyes and midnight-black hair lay across the back of a tapir with her cheek pressed to its ear, the water and mist cloaking her naked body as our boat swept past.

Sleep and consciousness alternated in dazed succession as the days and hours slipped into one another. It would take a week from Cuzco by air, land, and boat to reach the remote start of the expedition. In the solitude of these days, patience and fear, wonder and doubt jockeyed for favor. Soon the option of turning back would vanish. It felt as though everything had been reducing toward this moment, or perhaps building, my entire life.

For the final leg of the journey I hitched with a brazil nut farmer named Manuel and his wife. They were confused by the gringo in their midst. I told them where I was headed and they looked aghast, the way onlookers watch a drunk cross a street. They took me all the same after I traded a tank of gasoline and two hundred soles for the ride.

Santiago had warned that this was not a journey to take lightly, that getting to the Western Gate could be dangerous and past it all bets were off. Anything could happen. Thankfully, the fear I felt was counteracted by the anticipation, the thrill of the journey. I was heading into the last great wilderness—that shrinking thing that is vanishing from our wonderful planet; that world no one knows.

As the boat wound ever deeper into the jungle, Peter Beard's aerial photographs from Africa came to mind. Taken more than fifty years ago, his images show vast herds of hundreds, even thousands of elephants moving across the plains like ants—a vision of a much different earth just decades before my lifetime. There are no longer such herds of elephants; their ancient routes have been blocked by farms and roads. They have been poisoned by farmers and killed by poachers. The sad reality was that I had missed what Beard had the privilege to witness: Africa in much fuller glory than today. It struck me that I might be living a similar generational privilege in the Amazon, that the massive areas of uninterrupted and uninhabited forest might not exist a century from now. The weight of these thoughts, combined with the imminent wagering of my own life against the elements, nearly overpowered my determination to continue. Flashes of fear verging on panic would grip my throat, and there were moments when I almost told the driver to turn around. Yet as we continued to wind into seemingly endless unspoiled green, I was buoyed by subtle changes I saw in the landscape. The many hesitations my mind had concocted were continually washed away by the illuminating wonder my eyes beheld. The warm awe of witnessing such riotous biotic climax was hypnotic. It was impossible not to think about the significance of this place I was entering. My resolve was nurtured amid hours of meditation.

Ever since the La Torre anaconda expedition, I had been fascinated by the obvious negative correlation between human density and the abundance and diversity of a given area of forest. Traveling through the Madre de Dios, I had seen forest devastated by logging and hunting as well as places degraded by gold mining and its long-lived mercury contamination. But, in contrast, I had also seen the most secluded and pristine wilderness imaginable and the resulting riot of wildlife. Again and again the inverse rule held: the farther from humans you travel, the more animals you see. Now, traveling upriver on a boat with two strangers, I was considerably farther from civilization than I had ever been. With each bend in the river, the beaches were more crowded with tracks evidencing the visits of large animals, the hulks of basking caiman, and endless flocks of birds. Likewise, tree branches exploded with life. It was a celebration of human absence.

After La Torre and other expeditions, Pico and I discussed the inverse rule at length. We observed clear evidence that even sparse human presence can alter habitation of a given area of forest in negative ways. Even a lone farmer brings with him the din of a boat motor, the roar of a chainsaw, and the sharp report of a gun. He brings livestock, dogs, gasoline, fishing nets, light, and a plethora of other substances, sounds, and practices that have resounding effects on the surrounding flora and fauna. These and other observations we recorded spoke to the unique importance of untouched wilderness. During my subsequent visit home, however, I was surprised to find the idea of wilderness a point of intellectual contention. The very concept of the wilderness that now surrounded me was for some a matter of debate.

One line of reasoning sees wilderness as a myth, a modern construct invented in response to the wholesale vanishing of nature in the face of all that is human. In this view, wilderness is a romantic notion, a yearning for something that was neither labeled nor much appreciated when nature was omnipresent in our lives. Only when it had become apparent that wilderness was vanishing did people begin to appreciate, treasure, and worship an idealized version of the wild.

It is understandable that for many people words such as *pristine* and *wilderness* elicit uncomfortable reactions. Even many conservationists are uncomfortable with this language because it has the potential to devalue crucial and complex biodiversity areas in people's minds, and instead focus everything on places that better fit the "untouched" romantic image of wilderness. This is an increasingly salient point as many of the most important ecosystems and endangered species become intertwined with human roads, cities, farms, pipelines, and dams. Tigers and elephants in India fight for survival in a humanized, not wild, landscape.

Implicitly, this glorification of pristine wilderness denigrates nature as we know it, devaluing the more mundane and everyday nature of our backyards. Our distance from nature, in this view, undermines our humanness and contributes to our existential separation of individual from environment, in many cases justifying ecologically destructive behaviors. By extension, sustainability requires that we do a better job of understanding and acknowledging the importance of biodiversity and natural systems, even taking the rules of nature as a greater wisdom from which to derive our own conduct. Sitting in the front of a boat, watching the unimaginable verdancy around me, I realized that if humans must live in their natural surround, then it is a surround of compromised nature and not wilderness with which we must make peace. Such is the reality across most of the globe today. Yet, even with this proviso, it's not quite so easy in practice for us to achieve.

My experience in the Amazon told me that the argument against wilderness was both rational and agreeable and, at the same time, profoundly wrong. Indigenous cultures all over the world have sacred areas of wilderness where they do not live. Places they utilize for spiritual direction or rites of passage, or perhaps places that they do not enter at all. As I traveled further away from consensual nature and deeper into earth's "inner" (as opposed to "outer") space, a different realization unfolded within me. Humans must indeed learn to live on the earth that they know. In contrast, this earth that was swallowing me was not a place where most modern humans would be welcome or one they would understand. Here the rules were different. In this anachronistic wild, humans were not the top predator. It was a strange and novel vulnerability to experience. Yet even as the omnipresent power of the jungle stretched out, it was impossible to forget that the immensely powerful realm I now entered was at the same time so incredibly fragile. Anything but the most non-intrusive human presence could change and destroy it.

In light of this, the wilderness debate takes on tangible consequence. It may seem trivial at first to debate the specific syntax of words like *nature, wilderness,* or *pristine.* But in the Amazon these definitions carry genuine weight capable of determining the fate of entire species, cultures, and ecosystems.

For a long time, the wilderness discussion in Amazonia has become ensnared in the academic quicksand that it is the Amazon's human history (leave it to humans to make it all about us!). In 1971, Betty Meggers's book *Amazonia: Man and Culture in a Counterfeit Paradise* argued that the Amazon had historically thwarted any substantial human settlement and so was a pristine natural entity. Meggers saw no sustainable way to develop the Amazon and was largely influential in the "touch it and you'll destroy it" viewpoint that subsequently informed perception and policy.

In 1992, William Denevan published a paper, "The Pristine Myth: The Landscape of the Americas in 1492," which argued that Amazonia had indeed been inhabited by vast civilizations of ancient peoples. Some scientists even go so far as to claim that the diversity of species in Amazonia is somehow a by-product of past human disturbances, and that the biodiversity of the Amazon is man-made. Outside of the scientific community, current Internet sources claiming to debunk "eco-myths," go further to view the Amazon as "a purposefully engineered tree farm planted by humans thousands of years ago."

At some point in the debate, the prevailing message has morphed into the view that if humans had once been there, they should be there now—as though it would be irresponsible to let the gigantic forest "wasteland" continue to stand unpopulated by roads, bridges, bars, brothels, and industrial zones. Brian Kelly and Mark London bluntly ponder this view in *The Last Forest*, writing, "If an earlier civilization successfully settled here before Europeans came with their diseases and murderous ways, then why can't it happen again?"

So what is the truth? As of 2012, a team of leading Brazilian and American researchers published in the journal *Science* that, contrary to popular belief, "large ancient civilizations never cleared and tamed the western Amazon." The research, headed by Crystal McMichael of the University of New Hampshire in Durham, sampled soil from an area of more than three million square kilometers. The soil cores showed traces of charcoal and burnt grasses near river bluffs, but no indication of common Amazon crops. Inland from rivers the charcoal and other human indicators were far less common, further corroborating the long-held belief that most of the human presence in pre-Columbian Amazonia was along waterways. As for the spaces between, as one of the team stated, "If humans were in those areas, they didn't stay very long, and they didn't farm."

The team's findings are corroborated by the observations of the first European explorers to travel the basin. Their findings also fit with the hypothesis that without metal tools brought by Europeans, pre-Columbian people in Amazonia could not practice slash-and-burn agriculture and would have had relatively little ability to influence the forest.

In the end, when religions, political agendas, and the jousting of scientists are set aside to reveal facts, it seems obvious that whatever the exact nature of pre-Columbian civilization in Amazonia, there have always been wild places between *us*. There have always been great stretches of jungle straddling the horizon while generations came and went, where trees towered toward the sky, never once entering the human consciousness. In short, Amazonia is dominated by vast expanses of wilderness.

I use the word *wilderness* loosely to describe those areas

that are untouched, the shrinking part of reality crucial for the survival of species and the production of ecosystem services—among them, balm for the human psyche. But my fascination with wilderness goes beyond my love for adventure and wildlife, and my own personal, quasi-spiritual need for remote places. As a naturalist, I am interested in the baseline information that these inaccessible corners archive.

In New York and New Jersey, where I grew up, as well as in parts of the Amazon and India, I have witnessed a kind of generational amnesia to ecological abundance. It is a sinister phenomenon whereby members of each generation seem to accept what they see around them as the way things ought to be. It is a problem of shifting baselines, a lowering of the standards by which we judge the condition of our environment. Over generations and across continents, this collective inability to accurately assess environmental change has become a serious problem.

Growing up, I didn't look at the trees and wonder why the trees were so thin. And I didn't wonder why I was almost twenty before I spotted my first bald eagle over the Hudson River. It wasn't until I was older that I learned that the forests I grew up in had been clear-cut a century earlier and that bald eagles had been radically reduced by DDT only decades before my birth. What I accepted as normal would have seemed tragic to someone who had experienced the same places only a generation or two earlier.

This same narrative is playing out in different forms all over the world. In fishing communities, elders remember days of plenty we can barely comprehend. In India, I met people who recall great swaths of jungle filled with elephants and tigers; places that are now nothing more than a distant memory. I now began to wonder if something similar could be happening on a more subtle level right under my nose in the Madre de Dios.

Even in the limited time since my first journey to Las Piedras, I had witnessed a decline in the number of caiman, capybara, and turtles basking beside the rivers. Similarly, in the forest where JJ and Santiago's community of Infierno was located, there were many people who considered it normal that there were no herds of peccary in the forest, when in fact they had been hunted into local extinction only a decade or two before. I had seen these human-impacted areas, as well as the stark gradient of increased abundance and diversity that occur in the deeper, less accessible areas.

Therefore, as my solo expedition grew nearer by the second, my eyes scanned and studied the passing scenery endlessly. The gradients of impact were not hard to discern. As the boat traveled past the last deserted remnant of civilization, wildlife became far more abundant than anything I had seen before, save for the uppermost reaches of La Torre. Huge black caiman basked on the banks. Herons and lapwings steadily increased in number. Families of capybara became more common, hiding in the tall river cane. These places might never have heard the sharp echo of a rifle, or at least it remained an unfamiliar sound. It was clear here that animals used the land with greater abandon. With mounting excitement, I recorded these and other observations, scribbling in my notebook and sometimes testing out my camera.

More than once Manuel stopped his boat to hunt. He once hit a spider monkey with his shotgun, and then later the same day a peccary. Neither animal died quickly. I finished off the spider monkey with my machete, unable to watch his expressive face search for hope that would not come. Then I watched in helpless horror as the boar breathed through the bullet holes in its side, suffocating in a stream. How easily a life passes from a body, leaving nothing but cold, inanimate tissue. It was the last thing I needed to see.

I was eager to get away from humans and on with the solo, so when after so many days of travel on successively smaller rivers, and the canoe at last reached a point where we could go no farther, it was a relief. The jungle had been unbroken for two days of upriver travel, without a single sign of human habitation. This was the source of the river; the current was shallow and slow. Manuel smiled awkwardly in the heavy silence after the motor had been cut and said, "That's all."

I nodded and swung my pack out onto the large beach, trying not to think too much. "Are you sure about this?" he asked, squinting at me in the blazing afternoon light.

"Yes. I am sure," I told him.

"You are going farther upriver?" I told him I was.

"I have a map," I said, trying to reassure him, and myself. He asked if he could see it. To reach the Western Gate I would have to hike farther up the river we were on and then cross several miles of jungle to the Rio Moxos, where I would find the Gate. My plan was to travel for several days beyond the Gate and then cross yet again to a final river, the Jura, where I would open my raft and float down to civilization. The Brazil nut farmer and his wife scrutinized the map and illustrated route for several long moments before looking up, aghast.

"You cannot cross to the Rio Jura," he told me in Spanish with his index finger on the page. "There are *nativos* there— *calatos*—tribes. You'll die." He mimed shooting an arrow to illustrate his point. He was talking about uncontacted tribes.

"How do you know?" I asked, hoping his answer would be weak enough that I could still justify finishing the route I had been planning and preparing for over two years. I felt my temper flash for some reason, as though getting angry at this man trying to help me would do me any good.

"My brother is a fisherman," he said patiently, "and he fishes at the mouth of the Jura, where it meets the next river. Each year at the end of the dry season the nativos come to the beaches to collect *taricaya* [turtle] eggs on the upper parts. They will kill anyone who goes there."

His wife walked over then, glancing at the map in her husband's hands. "*Es verdad*—It's true," she whispered, her eyes showing their full whites in alarm. "They are very, very dangerous." He wasn't kidding. Encounters with uncontacted tribes very often end in tragedy.

The tribes that remain in isolation today have done so for centuries, even millennia. Their customs, beliefs, and laws are a world apart from ours, which makes them very difficult to interact with. One of the most chilling examples of this took place in Ecuador in 1956, when a group of American missionaries made contact with a tribe called the Auca. The Auca were well known for wanting to be left alone, and despite the tribe's history of violence, the missionaries spent time dropping gifts on a beach near an Auca settlement and waving from their plane. Over some weeks the missionaries interacted with the Auca people through hand gestures from the plane and the gifts they dropped from a box lowered on a rope. When they felt that the time was right, and the tribe seemed accepting, Jim Elliot, Ed McCully, Pete Fleming, Roger Youderian, and Nate Saint landed on what they named Palm Beach.

On the day of initial contact the missionaries interacted with the naked Auca, showing new items like balloons, shirts, insect repellent, and photographs from back home; they even gave one of the tribesmen a ride in their plane. But in their religious enthusiasm the missionaries were blind to the collision of worlds. As Wade Davis wrote, "The Auca had never seen anything two dimensional in their lives. They held the photographs and looked behind them to try and find the form of the image. Seeing nothing, they concluded that the portraits were calling cards from the devil." The corpses of all five missionaries were found days later, their plane reduced to shreds. Each had been speared to death for reasons known only to the Auca.

In the Madre de Dios, encounters with nomadic tribes have been similarly unpredictable. Even encounters that appear peaceful at first have become fatal, more than once resulting in deaths by so many arrows that the victims have been described as "porcupined" by them. It might seem harsh, but it is not for us to judge. The tribes have their own laws and beliefs, which we cannot understand. After all, the reason for their reclusion in the first place lies in the atrocities of the past, from the Spanish conquistadors, to the rubber barons, to loggers and gold miners. The very reason they survive today is their ferocious and uncompromising self-defense.

I had heard many firsthand accounts of tribes, many of them peaceful, from Don Santiago and others around the Madre de Dios. Other tribes were not. In 2011 a video filmed by tourists in Manu National Park shows a group of Mashco-Piro Indians walking on the beach. As the tourist's boat follows the naked men on the beach, camera rolling, the members of the tribe grow increasingly agitated and begin pointing their bows and even loosing arrows toward the tourists. Though ordinarily reclusive, the nomadic tribes of the Madre de Dios do not hesitate to defend what is theirs. In 2013 a group of almost eighty Piro Indians would invade the village of Puerto Nuevo, on the

Las Piedras River. Their hostile takeover was executed early in the morning, while the residents were still groggy from a night of festivities. The painted tribesmen gathered the women and children, pillaging houses and shredding beds. The terrified villagers could only watch as the Indians slaughtered livestock, taking machetes and other tools. It was a clear and threatening message: This is our forest, not yours. As the husband and wife looked over my map, I felt my courage and enthusiasm for my solo deflating rapidly.

They already thought I was *poco loco* for heading out on my own into the jungle, but it was evident from both of their gazes that now they were trying to determine whether I was suicidal. Taking the map from her husband's hand, the woman came to stand beside me. "*Escuchame*—Listen to me, if you stay on this river," she said, as she traced her finger along the Rio Moxos, "you should be fine. But if you cross over here"— now she dragged her finger along my highlighted route toward the Jura—"you will die."

There had been years of planning, and now, just as my solo was starting in earnest, the whole narrative changed. Glancing for a minute at the satellite image map, I searched for an alternate path. I mentally recited the mantra that I knew to be true after years of experience—that the plan falling apart is a natural part of any worthwhile adventure. If I continued up past the Western Gate, all the way, I could cross over the jungle ridge . . . Then if I went south for a bit . . . Within minutes, I had found a decent alternative to my original plan that would still allow me to make a loop while avoiding the Jura, but I still felt violated by the change. I wanted to get the possibility of tribes on my route out of my mind as quickly as possible. They were the last thing I wanted to have to worry about.

I promised the couple that I would not cross to the Jura and shook hands with them both. They gave me a lump of peccary meat wrapped in a plastic bag, which I stowed in my backpack. One last time Manuel asked if I would change my mind. "Come back with us," he offered, like he was trying to talk someone off the ledge of an apartment building. But I told him that I had to go.

They pushed the boat with a pole, half floating and half dragging against the sand. I watched as they moved down-river, becoming steadily smaller until I could barely see the orange of Manuel's backward baseball cap against the distant green of the forest. For a moment I felt a surge of panic and my hand twitched with the instinct to call out and stop them while I still had the chance. Yet as my pulse protested they were swallowed by the jungle.

From where I stood, it was well over a hundred miles of un-broken jungle to the nearest human settlement, weeks of travel on foot—days by river. This was it, like it or not. I was now a castaway in a sea of green.

16
The Western Gate

Here is sanctity which shames our religions, and reality which discredits our heroes. Here we find Nature to be the circumstance which dwarfs every other circumstance, and judges like a god all men that come to her.

—RALPH WALDO EMERSON

Idropped my pack on the beach and made some gear adjustments, placing the map back inside. I tried not to think about the fact that I had been dropped into the deepest wilderness on the planet and was following a dead man's tale with no map to a place no one else knew existed. Then again, that's exactly why I was here. Nerves be damned, I heaved my pack onto my shoulders, locked my jaw, and started walking northwest. When I reached a section of forest that looked dry and open, I spent three hours completing the first traverse, covering seven miles of jungle and emerging onto the Moxos for the first time.

When I had planned the expedition I hadn't expected to be dropped so far upriver. I was more than fifty miles ahead of schedule. From what I could calculate, I would reach the Western Gate in just hours. Although I had never thought about it, I realized I had been counting on a few days of solo travel before reaching the Gate to prepare myself mentally. I felt like a skydiver who has just taken the plunge with no idea if his chute will spring.

It took only two hours of walking before I rounded a bend and saw the tremendous gnarled column lying perpendicular across the river. It was just like Don Santiago had said it would be, unmistakable. I paused there, at the legendary boundary between the world of man and the nameless range beyond, suddenly timid. The Gate is actually a behemoth fallen tree laid across the river, with immense branches reaching up like a thirty-foot-tall defensive tarantula, a warning to those who would pass. It is a natural blockade across the entire channel, which marks the end of navigable river for boats. It's the last place with a name; the natural border between two worlds. For the local people in the area who knew of it, the Gate had spiritual significance: the beginning of the *otro mundo*, or other world. I stared at it in the dwindling light, humbled by its gravity. Crossing was not something to be done lightly. Even the fearless Elías had warned that beyond the Gate, all bets were off.

I made camp just below the tarantula of limbs, on the beach beside the river. Honestly, I was too spooked to cross in the dusk—even though "crossing" merely meant walking past it on the beach. My fear kept me below it that night, even though good firewood was scattered just feet beyond it, farther up the beach. It took a long time to fall asleep that night.

In the light of day it was far easier to feel the thrill as I took my first steps past the Gate. The sun crested the canopy around seven thirty and by nine the temperature rocketed up to a scalding ninety degrees Fahrenheit. I walked all day. The following day was much the same—seemingly endless trudging across beaches, interrupted only by hacking my way through stretches of jungle, on and on upriver.

I was astounded at how much my body was sweating during these days of travel. Shortly after starting each morning, my clothing was soaked with sweat so thoroughly that it looked like I had been swimming. I had to make a special effort to stay hydrated. In the Amazon, only a quarter of the rain is swept away in the flow of rivers, while roughly half the rainfall is sucked back into the sky through evapotranspiration. The vegetation drinks from the earth through billions of roots before being released once again through the trunk and the leaves; in the air the moisture accumulates to form clouds that once again dump the lifeblood of the system down to start the process over, so that the forest produces more than three-quarters of the precipitation it depends on. Forest and sky are one cycle. For this reason, when tropical forest is cut, the entire system is thrown off kilter—the drying caused by decreased vegetation interrupts the cycle and depletes vital moisture. Trudging up the beach, I, too, was engaged in this ancient cycle of liquid and gas, hydrogen and oxygen. Each time I would emerge from a concave bend of jungle to the open inferno of the beach, I would throw down my pack and dive into the river. I drank like crazy, with my chin and mouth submerged, just gulping in water. It was a wondrous gift after having grown up in a world where water is sold in plastic, or flows from metal pipes.

I drank in excess. I had been taught years earlier that the test of proper hydration in the wilderness was for urine to be "clear and copious," which it was. During an expedition, the body becomes almost like a machine, a system of functioning parts that needs to be deliberately maintained in working order: properly hydrated, lubricated, heated, cooled, fueled, and rested.

Around six o'clock the sun finally dropped below the trees and the temperature fell. I spent a few minutes filming some oro-cerulean macaws perched in a nearby tree, with the camera on the tripod. The brilliant birds chattered and nuzzled in the golden light as I watched them in the viewfinder, feeling profoundly thankful and full of purpose. This was exactly what I had come for.

That night beside a crackling fire, I checked my maps and saw that I had made some serious progress. While walking, I had kept count and recorded finishing upwards of twenty bends in the river, for a total of roughly eleven miles. A respectable distance. I jotted down notes on species and landmarks seen and marked progress on the map, occasionally swatting the lone mosquito. I boiled some water in a tiny pot and ate some ramen noodles, more because I knew I should than out of actual hunger. It felt spectacular. Things were going better than planned, and at this speed I would reach the headwaters of the river in two more days—the end of the Amazon and the start of the Andes.

On that night I pitched my tent in the middle of a sprawling beach. I had set it up only about ten feet from the water's edge, where the sand was slightly moister and held the stakes of the tent better. Inside with me were machete, headlamp, and a few other items. Outside the tent were backpack, paddle, packraft, and other equipment. I spent an hour taking notes in my field journal, writing and sketching before falling asleep.

Sometime in the night, a sound just in front of my tent woke me. The hairs on my neck quickly stood at attention—something big was nearby. Slowly I unzipped the front of the tent and folded back the nylon door. In the beam of my headlamp were two glowing red eyes not six feet from my face.

Most animals are equipped with a tapetum, a membrane in the back of the eyeball that bounces light back through the rods and cones of the retina a second time, allowing for greater vision. This is the reason for the "eyeshine" that most wildlife give off at night, and the reason they can see in the same jungle night that renders humans blind. If you spend enough time in the jungle at night you learn to identify species by the unique reflection of their tapeta in a headlamp beam. Cats have bluish-green tapeta, while many other mammals have white, and many fish have orange or silver. On snakes the eyeshine is constant and uninterrupted because they lack eyelids and never blink. On crocodilians the reflection is an unmistakable deep red.

Where the water met the beach lay a tremendous black caiman, the largest I had ever seen. She must have been upwards of fourteen feet. Her head alone was well over two feet long, her tail the girth of my waist, and from my ground-level perspective her teeth looked massive. I stared at her and she at me, neither of us moving. Looking left and right, I saw the eyes of numerous other large caiman, eleven in total. Some were as small as two feet long, others almost as large as the one in front of my tent—almost. Some were near the bank, others on the opposite bank, and some hovered mid-river.

The river before me was a swarm of red orbs. I knew I wasn't in danger: caiman are ambush predators, and if that big girl had wanted me, she would have done it already. I lay on my stomach, propped up by my elbows, staring at the crocs. Why there were so many crocs surrounding me, though, was beyond my comprehension. For a moment I wondered if I was dreaming—in my entire time in the Amazon, I had never seen so many caiman in one place. To find yourself alone in the wilderness with a host of giant saurians staring expectantly at you with glowing red eyes is a strange sensation, to say the least. They were *all* staring at me.

After several minutes of standoff, when neither of us broke the other's stare, the giant female turned her head toward the river and slid into the current. I ached to know what was going on. I closed the zipper nylon door again and lay down. *What the hell.* After several minutes of wondering if I should open the zipper again and check to see if the host of crocodiles had been a hallucination, I once again caught the sound of something large moving. This time I listened for some time. It was the sound of something massive dragging across the sand. With extreme caution not to make a single sound, I began opening the zipper. Peering through the narrow opening of a half-opened tent door, I turned on my headlamp. This time three large caiman were staring back at me. I unzipped the zipper all the way and waved my arms, yelling, "Get back!"

The large reptiles about-faced, and their broad tiled stomachs made that dragging sound as they retreated to the river. Standing up, I confronted them. "What the hell do you want?" I asked, now infuriatingly curious. I decided I had better move my tent farther back on the beach. I grabbed my backpack to move it and caught a foul odor coming from it.

The lightbulb went on: *that stupid peccary meat!* When I had parted ways with Don Manuel and his wife, they had insisted I take a piece. I told them I didn't need it, but they would not hear a rejection of their hospitality, so I had taken a small cut. I had done it just to appease them—I knew that the meat would go bad before I could eat it. I had stuffed the plastic bag with the meat into my backpack, intending to use it as fish bait later on, but had forgotten about it. Now it smelled terrible.

I looked back at the river with wide eyes of comprehension. The caiman had smelled the carrion and done what caiman do when there is a dead animal near the water: flock to feed. Taking the hunk of meat from the pouch, I began preparing for an experiment. This was the largest croc I had ever seen. I had to get her on film. I set up the tripod and camera ten feet from my tent and tied the meat to a piece of paracord, constructing a sliding trap. When the big croc came out I would reel in the meat, coaxing her farther and farther up the beach toward me. If it worked it would make for a great shot—face-to-face with a huge black caiman. The crocs kept their distance, cautiously waiting while I moved about. When everything was set, I climbed back into the tent.

The crocs watched cautiously and silently for forty-five minutes before the great female emerged onto the beach in front of me once again. First she came within two feet of the sand, then she placed her chin on the beach, then she took a first step; all at five-minute intervals. I was surprised to learn how cautious the largest predator in the Amazon can be. If she had wanted to, that croc could have sauntered up the beach, taken the meat, and had me as well, very easily.

After what seemed like forever, she finally approached the chunk of peccary meat, placing the front third of her body onto the sand and turning her head sideways to grab it. I pulled on the string and moved it just out of her reach. She righted her head and glared at me, obviously frustrated. I glanced over to the video camera and hoped it was capturing all of this through the darkness.

The croc took another slither up the beach only several feet from where I lay. Suddenly it felt too close for comfort. I was thinking of chasing her away again when she exploded into movement. She leapt to the front of my tent, and with her head turned to the right, decisively chomped the meat with a crack of monster jaws. Her massive mouth grabbed the meat as well as a scoop of sand—she also snagged the green mosquito net that had been rolled up outside my tent. She backed up as she threw her head back repeatedly, launching it all into her gullet. This had gotten out of control—I needed that net! I leapt up yelling, attempting to shoo her away with another round of shouting and arm-waving, when she took a threatening step toward me and roared. I fell back off my feet. She was swallowing my mosquito net whether I liked it or not. Everything went down her throat with unbelievable speed, and I was left holding a string that stretched across the sand and into the mouth of a giant black caiman. For a moment I held on to the rope, staring into her red eye in the beam of my headlamp. I reached as far forward as possible and cut the line. Satisfied, she turned to the water as dozens of other red eyes watched, the fat trunk of her spiked tail sweeping past before sinking into the river.

I was left shuddering. Part of me was thrilled over the encounter, but the ever-present question of survival took precedence: how the hell was I going to cook without a mosquito net? How was I going to peacefully observe wildlife without being constantly eaten alive? That mosquito net was a devastating loss to my gear. In the Amazon, bug bites accumulate to cause real problems like fever, infection, and severe mental stress. A lone human in the heart of the Amazon is nothing more than another mammal struggling to survive. I was a second-class citizen compared to what was out there. The only asset I had was the machete; that is the crux, the single most important piece of gear that exists, the twenty-something inches of steel that makes it all possible. Without a machete, life in the jungle is painful, difficult, and often short. Next on my list of crucial gear were my hunting knife, headlamp, pack-raft, tent, and then: mosquito net.

I knew that without that net, simple tasks like cooking, eating, taking notes, and filming would be made infinitely more difficult and discouraging. It worried me but I knew I had to concentrate on sleeping, to take advantage of the cool night hours. I had to sleep while I could, because at the crack of dawn, the heat and the insects would begin their siege. There were still a host of eyes watching me as I zipped my tent and lay down, knowing very well that it would be hours before I slept, if at all.

Indeed, the sun came up far too soon, and once again I broke camp and set out hiking before the bugs had a chance to find me. I took in the misty morning's beauty, pausing to enjoy the landscape, but by 8 A.M. it was so hot I knew that walking all day would be impossible. The sun was blinding and oppressive. I wished I had thought of adding sunglasses to the gear list.

I was still making excellent progress, and it was only day five of the expedition. It had been three days since I had left Manuel back at the Western Gate. Following the river, I didn't need to worry about getting lost at all. During the storm solo, it had been days of pounding stress and danger because of that constant risk of being irreversibly lost. It had been a stressful but important initiation for me. But this expedition, without that stress, was entirely different. It was fantastic. But I remained cautious when cutting through bends.

Cutting bends is something I do when the river cuts back so much on itself that it almost forms a circle. The rivers in the Madre de Dios slither to and fro across the landscape so languidly that bends in the river sometimes almost touch. In

places where this occurs, it is a much shorter distance to cut
through the jungle than to follow the contour of the loop.
Sometimes the distance is so short you can see through the
trees to the open space on the other side. Other times it may
be several hundred yards. The stretch of river I was on that
day had several bends that could be cut, and so I spent much
of the time walking under canopy, which was a refreshing
change. Stopping frequently, I filmed black skimmers, Spix's
guan, and an Amazon red squirrel. I decided not to film the
troop of squirrel monkeys that were playing above a small
stream since I already had good footage of the species and
needed to conserve battery. It was only day five, I reminded
myself.

As I walked I decided that the next day I was going to stay
put. The poachers had taken me farther than I thought I would
be able to hitch, and hiking, too, had been more successful
than I had anticipated. The jaguar tracks around my tent each
morning and the abundant broad prints on beaches made me
confident that if I was going to be able to film a jaguar anywhere, this was the place. I planned to set up a hide made of
palm and brush and sit in it all day, starting at 4 A.M. The
thought of sitting still all day without the mosquito net was
worrisome, and I wondered if it would even be possible. I wondered what the hell I had been thinking with that croc—I had
lost a vital piece of gear. *Should have had a backup for that,
add it to the notes for next time.*

Regardless, sitting quietly was the part I was most excited
for. Perched in the river cane, with an entire beach or perhaps the mouth of a stream in view, I planned to stake out for
days. In the jungle, where every single organism is eventually
eaten by something, every creature is on high alert, which is

why wildlife sightings in the rainforest are so difficult to come by, compared to other habitats. As humans, we usually scare away everything around us with our loud footfalls and obvious smell and appearance. But when you stop acting like a human and start acting like a part of the landscape, if you really just *sit*, you start to see the incredible world where animals just go about their lives.

This was what all the true wildlife film guys did. One of my favorites was Nick Gordon, a Brit who spent years in the Amazon filming, sometimes spending weeks in a single location, doggedly waiting for his subject. He filmed wild jaguars, diving with pink river dolphins and giant otters for the BBC and the National Geographic Society. In more than a decade filming in Amazonia, he documented the tarantula-eating Piaroa Indians and was once held captive by chanting Yanomami warriors. He hand-raised everything from spider monkeys to jaguar cubs, and even discovered a new species of primate. The guy was legit. Gordon had come from the other side of the world to find that he belonged in the Amazon, and he had determined to do whatever he could to protect it. He is one of those guys I wish I had had the chance to meet while he was alive, one of those guys I hoped to be one day.

His camera was his weapon. He went where no one else went, deep into Amazonia's shadows, and he filmed species in a way few ever have. Even today his work on jaguars is considered groundbreaking and has yet to be matched. The films he brought to the world generated international awareness and funding for conservation on a scale that is dizzying to conceive. He was a warrior who fought the bullets and chain saws with film. He defended the place he loved by sharing it with millions of people. And though I am not a professional cameraman, I knew that the footage I was shooting on this solo was something special. This place was special, unique on the earth, and if it was going to survive, someone needed to pick up where Gordon and the others like him had left off.

With my heavy backpack, slashing through the foliage between riverbeds, I daydreamed about the jaguar I was going to catch on film. I was still sore over the missing mosquito net, though. Nick Gordon lived the ultimate life but he had also paid the ultimate price for it. A rare form of malaria that he contracted from spider monkeys, along with the numerous other blows to his immune system that the jungle had dealt over the years, caused his heart to fail when he was fifty-one. He probably would have survived to be an old man if he had used a mosquito net more often.

I maintained a brisk pace without stopping for several miles. The jungle-beach transition came often. The river here was twisted tight across the land, searching the flat earth for a decline in elevation so that it could flow free. If you observe the lowland Amazon from the air, it is possible to see how the rivers weave like living snakes along the path of least resistance. From the smallest tributaries to the largest, they warp and wind through the jungle, carving out the clay. Over the course of hundreds and thousands of years their routes can change dramatically. As the rivers shift they leave behind scars in the soft land, sweeping curved lakes called oxbows.

Oxbow lakes are always near a river bend, a record of where the river once was, but they do not flow. Instead the sediment falls in the absence of current and an entire ecosystem emerges around them. Many species specialize in these stagnant sanctuaries; some species of birds, fish, and vegetation are found almost exclusively in oxbows. Black caiman specialize in ox-bow lakes, as do anacondas. But it was the call of a hoatzin that let me know I was passing the tip of a lake that day. The coughing call and clumsy wing beats are unmistakable. I de-cided to investigate.

It was a tiny lake, perhaps only fifty feet across, but long. The foliage was buzzing with a quiet energy. Squirrel mon-keys chattered in the low limbs above the water, where cai-man watched with stoic patience. Turtles basked on logs in the sun, and kingfishers zoomed back and forth; a train of butterflies flew with serpentine grace, slithering one after the other, winding through the foliage of the idyllic scene. It seemed like the perfect place to throw down my pack and spend some time relaxing, a good time to wait silently and see what came by.

I found a tree that grew at an angle from the bank and leaned over the water out into the middle of the lake. I climbed it and sprawled in its generous, moss-covered branches. Com-fortable and cool in the shade, the tree held me like a ham-mock. I watched the squirrel monkeys for some time, but didn't film—I wanted to savor the brilliant afternoon. Sunlight illu-minated the foliage, creating a green and yellow dappled roof above, birds chasing dragonflies, caiman, and turtles basking in the few spots ablaze with direct light.

I remembered Nick Gordon's story from *Wild Ama-zon* about filming a jaguar while he had been perched above a stream and the jaguar swam by. I felt similarly poised to witness such an event. How could any tapir or jaguar resist cooling off in the shallow pond? This was the fantastically beautiful setting I had imagined during my pre-expedition planning. Pushing my hat down over my face and folding my arms behind my head, I closed my eyes, dozing a little.

It couldn't have been more than ten minutes before I heard a substantial splash and my eyes snapped open beneath my hat. From the sound, it was clear that something big was coming. Turning slowly onto my stomach, I looked down over the lake, and in a halo of ripples were two giant river otters. They didn't see me; in fact they had no idea anyone was watching them. Cruising directly under me, the six-foot predators frolicked. The male would nudge the female and then flop onto his back and she'd chase, and so the two playfully swam for some time. It was the first time I had ever seen the endangered species, and they were significantly larger than I had pictured them. In fact, seeing them just feet from me, I got a surge of adrenaline—these were no-joke predators.

Their Spanish designation of *river wolves* is an accurate description. Throughout their range, giant otters live in packs and dominate their habitat. The longest member of the weasel family, their bodies are slender and powerful, equipped with large webbed feet, a powerful oar-shaped tail, and impressive canines. Their ears are close-cropped on a broad head and their snout bears a set of long white whiskers. The two below me had beautiful marbled patterns decorating their throats, and the dazzling rich coat of prime specimens.

I couldn't believe my luck. Most sightings of wildlife, especially in the jungle, are from the perspective of the frustrated naturalist as his subject flees into the bush—these guys had their guard down. My camera, of course, was in my pack on the ground—but I didn't mind. Otters are mesmerizing to watch. They seem to live life at 100 percent on, twenty-four hours a day, every moment of their lives.

The pair frolicked below me for some time, but then began to swim toward where my gear lay, at the base of the tree. Approaching within a few feet of my pack, the large male gave a sudden, violent snort that echoed through the forest, and the female mimicked him. Both dived into the water, disappearing. They surfaced nearly twenty feet back, lifting their heads high, craning to view the suspicious package that smelled so strange. From only ten feet away, and directly above, I watched as they circled and sniffed, snorting in alarm—they knew something they couldn't see was amiss. After half a minute I felt guilty and gave them a clue—a click of my tongue. Both heads snapped upward, eyes wide, barking aggressively and rising out of the water toward me. At the sound of their panic, another call of distress was voiced from an unknown creature—a long moan.

At the sound of that moan, the two parents raced to where they had first entered the lake and vanished into the foliage. I took advantage of their absence to rush down the tree to my pack and remove my camera. I had barely gotten it out of the bag when the parents returned, this time with a tiny juvenile between them. The camera was rolling as both parents raced toward me, barking and rising out of the water to display their white throats in an attempt to intimidate; the baby mimicked them.

Though both of the parents continued to scowl and bark at me from time to time, they eventually seemed to realize that I was not an immediate threat. They began swimming and playing once again, though always keeping a cautious eye on my position. They dived, and played and nuzzled each other as if purposefully showing off for the camera as I filmed. I watched them rip under the water and return to the surface, munching on fish. They clasped their catch in the same humanlike way a standing grizzly holds a salmon, with both hands. I was filming one of the most charismatic and endangered species in the Amazon, and they didn't mind my presence. In all likelihood they had never before encountered a human.

For some time they interacted together as a family, diving and curling in the water, nuzzling and grooming each other. The care and tenderness between them were striking; they were a family, going about their otter day in the jungle—checking for predators, catching fish, and performing other chores accomplished together as a unit. For a time they allowed me to observe their world, and threw little attention in my direction. When they moved toward the far end of the lake, I allowed them their privacy and carefully packed my camera away. For me, the rest of the afternoon was a continuation of beaches and jungle, beaches and jungle.

That night I camped as I had the previous nights. I gathered wood, lit a fire. Already, after just five nights of expedition, three days of solo travel, I felt as though I'd been out on my own for ages. I made notes about the otters and sketched some.

I was doing my best to hold on to my nerves and stay sharp on the expedition, but I did feel lonely. It is both exhilarating and cosmically terrifying to be so alone. The world I had grown up in seemed from a different lifetime: streets, cars, houses, toasters, cell phones, even other people. It is the most profound loneliness imaginable, as though the rest of the world had ceased to exist. I had a good laugh thinking of Times Square.

As I lay flat on the beach that night, the entire cosmos seemed visible: the lavender haze of the Milky Way, the faint dusty peppering of jade halos around multitudinous points of light and scores of more distant bodies. The celestial landscape exposed by the clear night was teeming with detail, great depths of the universe discernible. There were so many stars of such varied sizes that there seemed little room for the black of space.

As I stared at the stars hung at random in billions upon billions of miles of outer space, it suddenly meant something altogether different to be in the jungle. The Amazon, with its vast matrix of interconnected, interdependent organisms as seen from one of those points of light, a star billions of miles from earth, is something unfathomably unique. The Amazon, viewed from space, has been described as the Tree of Rivers: the trunk rising from the Atlantic Ocean and reaching across the continent in ever-diverging branches of tributaries. It is the tree of life, the single greatest example of life to exist, perhaps, in the universe—the miraculous antithesis to the trillions of miles of barren, frigid space.

Each mushroom, each decaying leaf is a world of its own, each a microcosm of fantastic complexity even apart from the whole they support. Yet there is no *apart* in such a world of cryptic complexity. The towering trees that rise from the barren clay of the basin support thousands of lichens, insects, mosses, fungi, birds, mammals, reptiles, and amphibians, and they cannot exist without the specific other life-forms of the biome. It is speciation in concert: each organism evolving alongside numerous others in the quest for life. For thousands of miles in every direction, as I lay on that beach, such was the landscape, vibrant with the riotous tumult of life.

It had taken considerable testing of my own mettle to make it so far—speaking plainly, I had been scared shitless for weeks. To journey into such dense and uncharted wild is to wager your own life; out in the jungle a lone human is just one broken limb, one wrong turn, and one miscalculation away from being absorbed and digested into nothingness. Many times I had thought of turning back, more than once physically doing so, yet always in the end pushing on.

Yes, the jungle, or any wilderness, is a treacherous place for an individual. But the Amazon is uniquely capable of swallowing a person, silently deleting them from existence. It wasn't just Percy Fawcett and the dozens who vanished behind him; the jungle swallows people constantly. JJ alone had told me numerous stories of locals vanishing. On the day before his thirty-first birthday, in 2000, biologist Francis Bossuyt walked to a lake near Cocha Cashu research station and never returned. Despite the ensuing weeks of searching by friends and colleagues, nothing of him was ever found; not a single clue to his fate was ever uncovered. Indigenous lore acknowledges the vanishings, suggesting that certain people are selected, inducted by the immortal beings in the stars to leave earth, and to do so without warning or trace. However you choose to explain it, there is an undeniable, awful tendency of the jungle to draw people into its bosom of shadows and remove them from living reality.

Although I carefully calculated every known aspect of my expedition, at any moment I could join Fawcett and all the others. In response to high-stress survival situations, the brain releases chemicals to create a state of peak experience, a heightened awareness to one's surroundings that I savored like a junkie alone on that beach. Each sensation, each molecule in my own body, seemed charged, connected to the landscape, immersed in something profound. With the humble fire crackling by my side, stranded on a nameless beach at the world's end, whatever might lie ahead of me was worth it.

For days I pushed on alone through the wilderness. At first

my skin was turned to raw meat from the constant wrath of the sun, and the siege of sand-fly bites. Usually in the forest there are 150 feet of canopy to guard against the harsh tropical rays, but traveling in the open beside the river was more like navigating a desert environment in terms of the constant dry heat. But eventually I felt my body adapting to the harsh conditions, and rising to the challenge. My skin began to harden and heal. Yet my mental transformation was most remarkable. My brain, too, was adjusting to the new reality of life alone in the wilderness, and operating with a level of focus I had never experienced before. Everything felt new.

Adventure in its purest form is raw discovery. The draw to see what's around the next bend becomes hypnotizing; I was drawn forward by the powerful tide of the forest. In the wild, the self becomes the only thing because there is nothing else. Out there the grave business of survival and discovery is understood without being spoken, and all else fails to qualify as relevant. Days of travel deepen this sensation, and you begin to feel the tug of the jungle, the sucking sensation of deep wilderness that draws you toward its hidden recesses. It was fascinating how powerful the urge to continue was, how much my mind obsessed over it. By that third night, my eyes felt wider, ears sharper, and gut more hollow than ever before. Reality was amplified. I was amazed at what my senses could interpret.

From where I sat, I could distinguish the odor of collared peccaries nearby, their musk more delicate than the white-lipped variety. The air bore the tangy odor of fermenting huicungo fruit, and the dank aroma of a blooming bride's-veil mushroom, ready to drop its skirt of fungal lace for the night. These interpretable scents were just notes amid the bouquet of various other aromatic compounds. In the insect din, the distant chanting of frogs, and the lonely tinamou's song I could hear the more subtle happenings of the forest. Somewhere upriver I heard the distinctive purring of pale-winged trumpeters, and farther off an Amazon horned owl. Without consciously thinking it, I knew that a predatory fish was hunting in the shallow water of the opposite bank, that a family of howler monkeys were bedding down for the night, and that capybara were preparing to emerge from the river cane nearby. I took pleasure in the chattering swarm of bats above my head, echolocating insects and snatching them on the wing; the sound of jaws crunching exoskeletons was constant and satisfying. I had front-row seats to the greatest show on earth.

There were behavioral differences in the wildlife as well. I saw species in daylight that would never dream of leaving the cover of dark back in the human-inhabited Amazon. There was no comparison between this isolated rangeland and the more familiar and accessible forests of the Madre de Dios. I spotted harpy eagles twice as often. And every morning there were jaguar tracks surrounding my tent. The heavy prints in the sand indicated mostly males, and one of them, a massive three-hundred-pounder, had boldly approached my tent during the night. He had come within twelve inches of the wall of my tent, the closest any of the jaguars had come while I slept. Another item to add to the "next time" list: camera trap. Kicking myself, I could only imagine the images I would have gotten had I brought one and faced it toward my tent.

Along with the jaguar tracks, the beaches were jam-packed with signs of other species, a Library of Congress of knowledge for me to record and decipher. On virtually every beach I found indications of tapir, ocelot, jaguar, agouti, paca, more than three dozen birds, tayra, giant otter, caiman, anaconda, and on and on. It was almost difficult to imagine so many species all using the same area. It was like a hidden Serengeti in the Amazon, a beach where every creature came to play, hunt, drink, and interact. This was what I knew I had to film.

But the discovery that gave me the greatest pause was undoubtedly a tremendous yellow-footed tortoise—the same species Lulu had attacked in our last days together, and the same one I had seen many times in the forest. However, this one was different. I spotted it when cutting a bend, and needed several minutes to compute what I saw. The tortoise was so huge that its carapace was almost three feet from end to end, and the animal's back was higher than my knee. It looked more like a Galapagos giant tortoise than the familiar yellow-footed one I knew. It was three times the size of the largest I had ever seen.

Kneeling beside the ancient reptile, I watched as its legs, heavily armored with thick scales, and wise, broad head were drawn into its immense shell. She hissed calmly. It had been decades since anything had threatened this creature. When it was young, luck had kept it from being crushed in the jaws of a jaguar like so many others of its species. Yet the time when a jaguar's gape could threaten it would have passed half a century earlier, long before the walking mountain of scale and bone had reached its final size. It was a stunning female, and I realized that no one I had ever met or spoken to had seen its equal—not even Don Santiago. With spine-tingling comprehension it dawned on me that surely I had reached a place alien to the world of men. The presence of the giant was proof: one of the most commonly hunted, easily found species had been left here to grow indeterminately, endlessly until it had become the mountainous dome of caramel gold I knelt beside. At that size no jaguar, anaconda, or even caiman could threaten the titan grazer. Surely well over a hundred years old, that tortoise must spend its days calmly plodding through the undergrowth, foraging, free from worry of predation.

I patted the huge dome of the tortoise and spent some moments in awed appreciation of the being before me. Walking on, I decided that soon I would stop, set up a permanent camp, and concentrate on filming for a few days. But for some reason I couldn't; my nerves wouldn't let me.

The forest was physically different here. The river water flowed free of sediment, bluish and pure—and colder, too. Here and there along the banks of the river, small waterfalls fell, cool blue highlights among the green. I was now far into the headwaters of the Moxos, the source of the river. This place was not accessible by road or plane or boat. It was a corner of the jungle hidden from the rest of the world, a true no-man's-land. The jungle here was dark and ominous, towering in ancient authority. Traveling onward the following day, I felt my courage buckle under the weight of the jungle's sacred depths. By day six I had gone farther than I had ever planned. As I walked quietly beneath the looming walls of jungle, the dark canopy snarled mist into the sky, and a creeping terror began to build within me. I no longer felt that this was a place where I should be.

The canopy loomed above, silent and dark, as if guarding some ancient secret. Rising out of the river were beckoning arms of fallen trees. Great vines hung from the emerald ceiling, tangling the dim corridor in warning. I walked in a state of constant, almost oppressive awe. It was as terrifying as it was fantastic. Every cell of my body seemed to sense that this was not a place I should be. On top of that gut sense was the growing, mechanical fear. I kept looking at the jagged debris of sticks and thorns jutting from the riverbed, and thinking of my raft. At this point I was now so deep in the forest that even the river's current would take days to bring me back to the world. If I hit a sharp stick, or was pulled over a submerged huicungo tree, that raft could be ripped in half. If that happened, I would have to retrace the progress I had made not only by hiking but through the remaining seventy-odd bends I had made by boat on the river. It was a dizzying distance to consider. And in such a remote area, it could be weeks or months before another boat traveled so deep. I began to feel uncomfortable. It also weighed on my mind that the Jura River, which Manuel had sworn was the home of tribes, was only a half dozen miles away.

After days of hiking and camping, watching and listening in the silence of the forest, I felt a crushing sense of isolation, and a deep loneliness settled in my chest. The thought that it would be weeks before I would see another human hung eerily over me. Even the softest footstep on the sand echoed loudly here. I felt like an intruder and could only pray that this hallowed ground would grant me passage.

The day was silent, still. I tried talking softly to myself just to hear a human voice, but talking to me was boring, so instead I had an imaginary conversation with Steve Irwin. If anyone could lighten a situation and lessen the fear, it was Steve. He was no stranger to being out in the wild alone. After all, the guy had lived in the Australian bush for years by himself, catching crocs with just his dog and a boat. He knew what it was like to be on his own, and anyone who has been out there alone, whether in the Arctic or the Amazon, knows that the mind can dwell in dark places. But was I freaking out for no reason or was my gut trying to tell me something? I couldn't tell.

My conversation with Steve went on for some time. I told him about the wildlife I saw, the crocs, the otters, and he repeatedly responded with an enthusiastic "Crikey!" I vocalized the entire dialogue. Steve's voice was full of all the encouragement he undoubtedly would have offered had the conversation been real. It might sound comical, but the talk with Irwin was actually helping, until the thought of his death entered my head. He had been killed by a stingray, a usually harmless species, while doing an ordinary day's filming—a mundane day at the office for him. Even the invincible, fearless Irwin had been struck down in broad daylight before he had the chance to do anything about it. My brain seemed set on finding pessimism no matter where I sought reassurance. That fact in itself was worrisome to me. I wondered if I shouldn't just make camp and spend the day staking out a beach, hoping for a jaguar—but I was feeling too uneasy, too nervous to stop.

Ten A.M. found me walking through jungle, atop a steep cliff where the river drove into the bank below. As always there was a beach on the opposite side of the river, a long one. The view from the elevated vantage point was stunning, just wild Amazon all around. "This is crazy!" I shouted to the jungle, and it echoed in the silence. I gave a long yell to the wind and hit a fist against my chest. It really felt crazy; after days of being alone, I was starting to wonder how rational my decision to journey alone had been.

I was determined to push past the funk. A light rain began to fall. In the soft hiss of the shower the journey continued in silence, but I hadn't gone ten yards before my eyes were pulled upriver by a sight that drained the blood from my face. Rising from the canopy on the opposite side of the river was a narrow column of smoke. From where I stood it was impossible to see where it came from. Dropping my pack, I hustled through the forest toward where I could see the source of the smoke and crouched down, suddenly regretting having screamed minutes earlier. I made my way forward along my side of the river, slowly expanding my view of the bend in the opposite beach, until I saw them.

17
At World's End

Have you journeyed to the springs of the sea
or walked in the recesses of the deep?
Have the gates of death been shown to you?

—Job 38:16–17

Nestled in the center of the expansive white beach were three small palm huts. To the left, farther upstream, a human shape crouched and vanished in a rustling of foliage. Three naked men moved in the open just under the river cane. One walked in front, cautiously but with confidence, beside the tree line. His right hand gripped a long bow, with arrows clasped to the shaft, and even from across the river I could see the foot-long bamboo arrowhead tips. He was the only one I could see clearly, and he held my gaze—I could not tell how many there were altogether.

I wished that I were hallucinating, but I wasn't. I just wanted to wake up. *What were the tribes doing on this river?* The beach where the men stood was littered with shavings from branches, imprints of recently made footprints, and there seemed to be something cooking on top of the fire. There was no question that they had heard me yelling moments earlier and now were taking cover in the brush.

My expedition was over. My life, too, was possibly over.

Their elders would have educated them to the fact that most outsiders mean trouble, even death, and they'd be justified to defend themselves. These were people who did not speak English, did not speak Spanish; they had never seen a wheel, a wrench, a spoon. They had never been inside a school, listened to the Beatles, or seen a car. They had never heard of World War II, they had never heard of the United States, hell, they had never heard of the country they lived in. These people were literally from a different world. Their beliefs, values, ambitions, fears, and motivations were made up of elements I could not understand, communicated in a language that no one from my world knew. It was possible that the people living in those huts were just as scared of me as I was of them; it was also possible that they saw me as a threat that needed to be hunted and removed. There was no way to tell. No way in hell.

At that moment, as I stared at the green palm that made their dwellings, arranged in three small huts, my heart sank. I remembered the peccary that Manuel had shot days earlier, fighting for air with its mouth in the water, lungs punctured. I remembered the spider monkey, so full of life and the desire to live. They had died before my eyes. The sound of their death, the sight of watching the life pass from their eyes just days earlier had stayed with me. It isn't difficult to die; in fact it is very easy. As I stared at those huts, my own mortality was so tangible it made me sick. Actually, *sick* is not the word; the sensation I felt was worse; I could taste my own death. This was too far. This was the answer to the question I had pushed and clawed at for years—always yearning to plunge deeper and deeper into the jungle's most forbidden depths. I had tested the boundaries and at last found the limit.

For maybe a minute I stood frozen, just watching the beach, looking into the shadow of the forest to where the one human figure strode slowly, watching me. I could see the outline of his muscular, naked body, but his features were obscured by the distance and shadow. The other members of the tribe were mere shadows in the bush, but I could see them peering curiously back at me. I remained where I stood, waiting to see what would happen next. Would they approach? Would they try to communicate? Would they attack? *How far could that arrow fly?* For some reason my hand twitched, instinctively wanting to wave to them, to show a sign of peace. I couldn't move.

Thrumming behind my fear for my own life was another fear—even on the chance that these people weren't violent, *my* presence could be lethal to *them*. Too many times in the past, hundreds if not thousands of natives living in isolation with few antibodies have been killed by the exotic pathogens of outsiders. Everything about this situation was out of control.

There was no plan for this, there was no one to consult, no previous knowledge to draw on; encounters of this kind are virtually absent from history in the last century. In a moment my expedition had gone from living the dream to being a nightmare. What I experienced at that moment was such devastating fear and regret that it took me weeks to suppress it. *Please don't let me die.*

I watched the man walking toward me, only a hundred yards now from where I stood, and realized that in all likelihood I was about to have a unique encounter. What could I do if they started crossing the river? What would they be like face-to-face? I was trying to think rationally, but instead was just panicking. All I could think was how to possibly get away. The only way to describe the sensation is that it felt like death. Under stern clouds a moment suspended in heartbeats endured.

Turning, I strode for downriver, then broke into a run. Sprinting through the jungle, I dodged trees and vines, thorns and streams as I made flight. In terms of body language, I figured nothing could be a clearer indication of submissiveness than running for my life. I was terrified on so many levels that all I could do to fix the situation was to sprint with all the strength I possessed.

For an hour I ran without stopping. But I began to realize that I was doing something very foolish—draining my body of every ounce of energy I had, and this had become a survival situation. Gasping for air and shuddering, I exploded onto a beach and ripped open my backpack, removing the most precious piece of gear I carried: my pack-raft.

Throwing the raft open, I took out the inflation bag and began filling it with air, then squeezing the air into the raft—all the while snapping my head left and right, looking for signs of humans. *This is how it fucking happens*, I hissed to myself as my hands shakily fit the inflation bag into the raft. *This is how people vanish—they run into the last thing they ever dreamed of and it kills them, and the animals eat their body before dawn the next day.* I had always wondered if people knew when they were screwed. Was there a moment of comprehension as the jaws dragged them beneath the water, or as the arrow parted ribs, or as the massive tree toppled to the ground?

Using the inflation bag, I captured large quantities of air, and then hugged it close. The air from a single bag was more than a dozen lungfuls, and after just five bags the raft seemed workable. I snapped the sections of paddle together, grabbed my pack, and climbed in. The Moxos bore me downstream with a speed I could never have matched on foot, and I paddled to add to it, suppressing the projections of my mind in which an arrow tore through the body of the raft—*pssshhh!* But no arrows fell.

On the river, however, I quickly had a new set of factors to worry about. The current was swift, nearly ten miles per hour and much faster than I would have imagined, and the river was full of debris. Carcasses of giant trees reached out of the water, thorny limbs lurked in wait, and floating logs threatened to pop my raft. Zipping past huge logs, around palisadas, the nimble little craft did its part and I did mine by steering and keeping wide eyes for the safest passage. Pico's many teachings came into play big-time here. The ability to read the river, to know where there was something beneath the surface or something too narrow to fit through, was crucial. That day, for the first time, the gravity of my situation came into full view from wherever it had been hiding: if that raft popped, it truly would be months before I found my way out of the jungle.

The river swung me around a bend and an island in the middle forced me to choose left or right. I chose right, where the channel was larger and the current faster—to the left looked shallow and full of dangerous, sharp debris. As I maneuvered the raft to the right, the current picked up even more and I braced as the raft plunged over a drop of two and then four feet. Thankfully, the declines were gradual and the raft took them in stride. I was able to cover in one hour what would have taken six walking. Even if there were tribal men chasing me with bows, I knew there was no way they could keep up with the speed of the river without canoes.*

Moving swiftly through the land, under a dark sky, the river demanded so much of my concentration that there was no room left to worry about the tribe. Even a moment's pause from the constant scanning and calculating could result in a collision that could tear the raft and plunge me into the river, stranding me in the jungle. I passed under the towering mountains of debris, everything from tree trunks to twigs piled thirty feet high over decades of flooding and receding.

Traveling past one especially large mountain of fallen trees and branches, I startled a ten-foot black caiman that had been basking on the beach. The croc burst into a sprint toward the water and in a fraction of a second came rocketing across the mud toward my raft. Our trajectories were perpendicular and set to collide. I screamed, and braced for impact as the startled reptile reached the bank and dived into the water. Speeding below me, its back crashed into the bottom of the raft, and the raft nearly overturned. But remarkably no damage was done. The tremendous scaled body vanished beneath the raft and into the river. It had only been trying to retreat to safety; I had just happened to be in its way.

When the current finally reached a straightaway, I breathed a huge sigh of relief and flopped back to lie on the raft. I dipped my hand into the water and splashed some water on my face, then drank. With a sweep of the paddle, the raft easily spun 180 degrees; setting the paddle down I let the raft float downstream backward, watching upriver in paranoia. Thankfully,

---

* Although many voluntarily isolated indigenous tribes do use various types of watercraft, usually burn-hollowed logs, the tribes in the Madre de Dios seem not to do so. In the dozens of personal accounts I have noted over the years, no one who has seen the tribes has ever observed them with a canoe or boat.

there was nothing to be seen except for quiet jungle, and, turning downstream, I continued on. That day I covered more than thirty miles of winding, downriver current. I basked in the fact that there was no way any human could keep up on foot with the sustained pace of the river, and once again I was safe.

Traveling downriver was freedom. For the first time in hours, I thought that I might survive the day. For the first time, I could be angry.

I cursed bitterly, furious that so many months of planning had just gone down the drain—so much for sitting in a hide and filming wildlife, so much for finishing the 157-mile route. That was all gone now. I also knew that no matter what, after an encounter like that there was no way that I was sleeping peacefully again on this river; that ruled out continuing the expedition farther downstream. Yet in those hours of pounding anguish and fear I discovered something that would change my whole future relationship with the jungle.

In the years I had spent in the Amazon, I had never before paddled alone on a river in silence. Most of my travel had been done with JJ and his brothers, traveling as the locals do in roaring boats that destroy the silence and scare every creature for miles around. Even aside from the incredible noise, the machine-propelled, thirty-foot canoe made it impossible to feel the river as I did on that day.

My legs could feel the cool, churning current below the raft's bottom, and through my paddle I seemed to be able to feel each rush and eddy. In the raft, small and silent, a new world was suddenly accessible. It was so much more pleasant than the endless battle of trudging upriver on foot; it was indeed the most serene experience I have ever had. The swift current and my careful paddling brought me to within feet of capped heron and then a sun bittern, a family of capybara; one time a huge stingray glided beside me for a while in shallow water. Black skimmers, a species of bird with an elongated lower beak made for catching fish on the fly, darted around me, picking off prey from the river. Once, while passing the mouth of a stream, looking right, I caught a young tapir eyeing me curiously. Twice I spotted medium-sized anacondas basking on the banks. Several times I startled legions of butterflies that had been lapping salt traces on the beach; they rose— thousands of them, dozens of different species—into brilliant vortexes of every conceivable color. It was surreal. Rafting that day changed forever the way I interacted with the jungle and the Amazon.

I rafted the entire day. Sundown came around 6 P.M. and I pushed on for another hour with my headlamp. Altogether I rafted for nearly ten hours that day, covering well over fifty miles before finally making camp.

Never before had sleep seemed so daunting. How could I zip my tent and turn off? The answer was, I couldn't. For hours upon hours I lay in the dark, listening for every sound. It was impossible to tell if I was awake or sleeping after a time, but I would check my watch to see that an hour had passed, or fifteen minutes. I know that I was sleeping at least once because I woke up with a start, I could hear whispers. Heart jackhammering, I realized that there were voices in every direction surrounding the tent. *Holy shit!*

As an arrow penetrated the wall of the tent I awoke from the dream with sweat pouring out of me, panting. I unzipped the door and stood on the beach, scanning right and left. All was still; no one was there. I wanted to keep rafting but knew it was too dangerous in the night—the raft could easily hit something and pop. I knew that in all likelihood the tribe was miles away and I *should* have been fine, but anytime I drifted to sleep I just saw the painted faces of warrior Indians attacking, or stalking up toward my tent. When awake, I was heartbroken. So much planning and anticipation, so much had gone into this expedition, and now it was over before it had begun. The weeks of filming I had planned to do, the adventures I would have had—all of it ruined. But there was no way I could stay on this river knowing that I might be putting myself and the tribe in mortal danger.

The following morning I rose before dawn, inflated my raft, and pushed out once again. The sun emerged on the horizon and ascended into the blue, cloudless sky. Many hours passed as the river whisked my craft through the jungle. The speed varied, sometimes as fast as fifteen miles per hour and other times, at wide sections, slowing to one or two. Most often, it was somewhere around ten, and over the course of the two days, I estimated that I had journeyed between sixty and eighty miles.

By 2 P.M. on the second day the river began to change. I had made serious progress downriver, retracing all the ground I had covered on foot and more. Now I entered an area that I had not noticed on the way up. The topography was stunning. Another sweeping turn of the river bore the raft in its current, and I watched as the jungle rose high above. The bank to the right was almost vertical, rising more than seventy feet in a sheer wall of clay and scrub before giving way to jungle. At the top of the steep cliff the roots from ancient trees stood naked in the wind, exposed by the fallen earth. I paddled disinterestedly as my eyes remained fixed on the view above. The turn ended and the river became a straightaway. It was then I spotted a boat, then another. Inside the boats were four men.

My heart quickened at the sight of humans from my world, and an indescribable primal elation flowed through my veins at seeing other people. My raft approached the two boats and I dug the paddle into the current on the left to turn sharply onto the sand. The four men stared at me, eyes bulging out as I came by in the raft. We talked for several minutes and I learned they had come as far up the Moxos as they could, looking for good hunting—and the last thing they expected was to find a white guy. It was also revealed to me that one of the party, a fat cross-eyed member of the smaller vessel, had just shot a jaguar.

"What?" I asked, hoping that I had heard wrong.

"Sí!" said the cross-eyed man, grinning broadly. He was short and fat, with a dim expression. A filthy baseball cap covered his ratty, balding hair and matched his soiled and unbuttoned shirt. The second man, whose face I couldn't see, was skinny and silent, uninterested in the conversation. I scanned both boats and saw no spotted cats among the many carcasses there. The man explained in Spanish, "We shot it just a hundred meters upriver, high up on that fallen log." I knew exactly where he had meant. Just minutes earlier I had made a mental note of an impressive log at the top of one of the cliffs. It had occurred to me that it was the perfect perch for a jaguar to survey the river below.

"You hit it?" I asked.

"Yes, but then it ran into a cave."

"So you are just going to leave it?"

"Sí! A wounded jaguar is very dangerous." At this point it was difficult to control my rage. I tried to explain that you can't just shoot something and leave it, least of all a jaguar, but the men only grinned more and shook their heads. I looked upriver, to where the last bend was, just beyond where it was possible to see; there could be a wounded jaguar there . . . a jaguar.

"Tu tengo miedo?" I asked. You are afraid?

"No!" said one. "Are you prepared to go looking for it up there?" I told him I was. I was feeling violent with emotion. "Ta bien," he said, and motioned for his partner to start their motor.

As we covered the hundred meters upriver to the place of the incident, the men filled me in on the details. Somewhere along the way, I gleaned that the chattier of the two poachers called himself Lucco. Lucco explained to me that the jaguar had been sitting on the large fallen beam that lay across the cliff above before they had shot it. Looking up at the large beam, I could picture the large cat stretched out there above the rest of the jungle, regally surveying its kingdom. The setting was perfect, exactly the place a person would expect a jaguar to spend its morning, any photographer's dream. From his description of the jaguar's initial position he seamlessly continued to the part where he raised his gun and shot the animal. From the log, he said, the jaguar had fallen and then quickly regained its wits and disappeared into a cave; as he talked, my head spun. What's more, I had no idea what he meant when he said "cave." Caves are rare in the homogeneous clay landscape of the western Amazon.

The boats were tied. Lucco opened his 16-gauge and removed the expended cartridge, throwing the red plastic and metal case into the river. Then he asked me to bring my headlamp. The bank was steep and so slippery that we were forced to use all fours and still found ourselves at risk of being thrust downward into the river. While we were climbing, I repeatedly asked Lucco if he was certain his shot had hit its mark. I asked repeatedly because I just couldn't fathom someone actually raising a gun to a jaguar, it went against everything I knew about local people's supposed respect for the cats.

The other thought I had was one of self-preservation: if we really were about to come close to a wounded jaguar, the situation needed some considering. With each moment we climbed higher, my heart began to pound ever harder. The cliff was so steep that we had to kick at the mud to force our toes to hold, and sweat poured from every pore in my body as repeated slips threatened a long and painful fall. Each time the mud gave, I froze, clutching the cliff with every bit of strength I had left, machete in my teeth. As we scrambled skyward, our legs and arms were repeatedly snatched by razor grass that tore our clothing and left deep lacerations on our skin. After several minutes the boats had become small beneath us and the incline mellowed so that we were able to stand. We were beneath the log.

Lucco brought his gun into position and gave me a grave nod. Standing several feet from him, I clutched my machete. It was clear now what he had meant when he said cave. Beneath the cascade of roots and limbs from the jungle above were large hollows in the upper area of the cliff. It was difficult to tell how deep they were from all the interference, but the sight of those black chambers beneath hundreds-of-years-old trees tugged at my gut. There was no turning back now. Back down by the riverbank, when the men had first told their story, I had processed it as just that—a story, a collection of descriptive statements and little more. Now, as I stood at the top of the cliff, faced with the reality of our position, the sight of those caves, my skepticism was the first thing to fall away.

At that moment, I knew there were only three possible outcomes to the next five minutes of my life. One, of course, was that we would see no sign of a jaguar and return quietly to our boat. Another was that we were about to be faced with the savage wrath of a wounded jaguar defending itself against its attackers. The third and most terrifying option was that I would have to watch as the already wounded jaguar was executed.

We advanced slowly, Lucco with his gun outstretched in front of him, ready to shoot at the first sign of anything. It was then that I began to feel truly sick. Even as I veered between fear of an attack and dread of what I might see, my eyes clung to Lucco as his trembling legs brought him forward. I hated that man. *You bastard*, I thought, *you miserable murdering bastard fuck*. I found myself imagining a jaguar tearing out of the cave at full tilt, ears peeled back and face contorted with savage wrath. I pictured the cat's broad paw clubbing Lucco in his fat poacher face or, better yet, removing his throat in one slash. I hoped that it would happen. In fact I was rooting for it.

My blood-soaked daydream came to an end as we reached the mouth of the first cave. Lucco hissed for me to turn on my headlamp, which I did. He parted the hanging roots with the muzzle of his gun, peering into the dark. I prayed not to see the cat. I prayed that Lucco had missed, and that the jaguar was a half mile away by then. It was all too easy to picture the wounded cat huddled in the shadow of the cave, helpless, and Lucco mercilessly executing it at point-blank range. *Please, no,* I repeated to myself again and again as I shone the light into the reaches of the small cavern. There was nothing.

The first cave was indeed empty, and so we moved to the left to check another. When that cave also turned out to be vacant, Lucco seemed to relax a little. He looked my way and shrugged his shoulders dumbly, now scanning the ground for tracks and still holding his gun at the ready. There were no signs of the jaguar's impact after falling from the log—it was as if it had simply evaporated. I too scanned the clay for tracks. Jaguar tracks are large and telling, but we saw none.

After several tense minutes of watching and listening it seemed as though the cat was long gone. Lucco lowered his gun, moved off farther to the left, and mumbled that he would move to check other caves off to the right, while I remained where I stood. I tried to reconstruct the event, and deduce possible trajectories a wounded animal could have taken when falling or leaping from the log. It seemed that with so much soft earth, there had to be some evidence. I re-created the event in my head as Lucco had told it, and imagined the cat being shot and falling from the log. If it had fallen injured there would be a large, smooth impact crater. The log was more than ten feet from the ground, more than high enough to produce a noticeable impact. That crater should be more or less just below the log—where both Lucco and I had already searched. Looking at the boats below, and then considering Lucco's pathetic antique excuse for a gun, I began to wonder if he had—or could have—actually hit the cat from that range. It was quite a distance, and the pellets of the 16-gauge might not have held their power by the time they reached their mark—if they reached at all. The thought dawned on me that it was possible that the cat had not fallen from the limb but instead jumped in escape.

I broadened the radius of my search, meticulously inspecting every inch of earth. It was then that I saw them: four large, broad jaguar tracks punched deep into the mud, with several large spots of blood beside them. The blood beside the four paw prints in the clay settled the debate: Lucco had shot a jaguar.

The tracks were set in the clay more than fifteen feet from the log. My tracing of the trajectory from takeoff to landing revealed that the cat had taken a massive leap from the downed tree in which it had been perched. The distance covered was considerable; the beast in our vicinity had to be dangerously athletic. My heart was racing.

Kneeling by the impact area, I looked from the log to the prints and then ahead. There was another cave, one that hadn't been checked. The moment my eyes fell on it, I knew the jaguar was inside. It made perfect sense, given the evidence, and the earth does not record such things with error. The jaguar had bounded from its perch, landed, and then leapt directly into the shelter of the cave with barely a trace, as if it had never existed.

There was nothing I could do; one foot went in front of the other. There was no chance of turning my body around, although areas of my mind were rioting in fear. It was just another instance of wet paint and my need to touch the wall; I knew I'd regret it the rest of my life if I didn't find out. Reality moved in dream time as I approached the curtain of roots and foliage at the mouth of the cave and knelt. I cautiously stretched my arm forward to hold the curtain aside. Light flooded into the darkness, illuminating two large golden jaguar eyes, which stared back into mine.

All noise stopped. Every sensation and all awareness of a world existing beyond the grip of those hypnotizing eyes vanished. Not three feet from my face, and only inches from my outstretched hand, was a fully grown jaguar. Her eyes were wide, wild, livid with betrayal—with such savage intensity that it felt like her gaze physically penetrated my head. At the center of each eye was a pitch-black sphere surrounded by a starburst of green and orange. She was panting, mouth ajar so that just the bases of her two large lower canines were visible against the black of her lip.

My hand was only a foot from her face. The indescribable kinetic power coiled in every inch of her body dwarfed the defenseless creature before her—me. From the impressive weight of her broad paws, where her retractable claws were hidden, came the muscled pillars of her forelegs, followed by a barrel chest and heavy neck, above them pointed shoulder blades; even the war-paint rosettes that decorated her coat were testament to her supremacy in the animal kingdom.

I knew I was in extreme danger. This was an animal that could run faster, jump higher, think better, and be far more lethal than any human could ever dream of in this setting. Our two bodies were separated by only air; whether I lived or died was entirely at the discretion of the cat.

Yet even overlooking the mechanical advantages of the jaguar, there was such a fire in her, a wrathful vibrancy of life emanating from her, that my own fascination overpowered the instinct to flee. She seemed to exist on an entirely different plane of consciousness, immersed in a constant state of peak experience unobtainable to the ever-distracted human mind. Looking into those eyes was as overwhelming as it was intoxicating. This was a creature that, except for the brief days in safety as a kitten, swatting at her mother's tail and taking in a new world, had spent every waking moment fighting for life. Like one endless solo expedition, her life was a solitary battle for survival. Each morsel of food won required skill, stealth, and often the risk of injury or death. Stalking the beaches and brush of the long river beneath black thunderheads of the Amazon sky, she had managed to survive through all adversity in one of the most competitive and unforgiving ecosystems on earth.

After several seconds I noticed that I was still alive. . . . She wasn't attacking; instead she was just staring back at me, panting. It was almost impossible to believe at that point. Had I been able to move I would have run, but she held me with her stare.

She looked desperate, hunted. Even in those blazing seconds, the beauty of that face struck me. Her features had the lean aesthetic of a female, almost leopard-like. Between fear and wonder it was nearly impossible to break her gaze, but the will to know forced my own eyes to dart for an instant to her flank. The pattern of her fur was not the small markings of some jaguars, but broad rosettes, almost like a Borneo clouded leopard. Following her pattern, I saw a stream of blood running toward the ground, where it collected in a small pool. She was wounded. When I saw the blood, I felt my pulse intensify, and a new sensation overcame me, one of complete despair. There was no way to discern the extent of her wounds, but from the amount of blood, it was clear that her injuries were serious. My heart sank. There was no more hoping. Lucco had hit his mark. In all likelihood the jaguar before me, the fantastically beautiful ruler of the jungle, would die a long, slow death from blood loss or infection—for nothing.

This animal was looking directly into my eyes with no indication of aggression, only lacerating desperation, pleading and imploring the stranger before her to simply let her be. I'm sure that more than a few people reading these words would accuse me of anthropomorphizing the cat, or romanticizing her communication, but what I saw was unmistakable. We all can see when a dog is bored, or happy, or confused; we all have seen the big cats at the zoo with that empty, spiritless glaze in their eyes from a lifetime of caged misery. This was no different. The jaguar was communicating. Terrified and badly wounded, she was doing all she could to simply avoid further persecution. In the space between us there was no escaping the untold atrocities inflicted by our species on hers.

When the sound of Lucco crashing through the brush reached my ears, I glanced left. The jaguar's jowls curled in a silent snarl, exposing tremendous yellow fangs for a moment, before once again falling. My heart was slamming in my chest; if he came to know she was here, he would surely kill her. Now panic coursed through every inch of my nerves. If Lucco discovered the wounded cat, he would finish her off in the cave, where she was helpless. The fantastic fire in those eyes would be extinguished.

His clumsy trajectory was leading him right toward the cave, where the jaguar hid and where I knelt, and if he continued he would almost certainly see the jaguar tracks, large and clearly defined in the clay. Once more I met her eyes, so unbelievably close to my own, but there was no time. I lowered the curtain of roots and brush that I had been holding aside. "Please live," I whispered, holding her gaze before the last glimpse of her disappeared behind the brush. Cautiously I stepped backward, ever aware of her ability to change her mind and choose fight over flight, and moved away from the cave.

Lucco was now making his way toward me, swearing in Spanish. As he slipped and fought his way over the steep landscape, my eyes went to the large pawprints still in the mud. *Shit!* It was miraculous that Lucco hadn't seen them on first inspection, and there was no chance he'd miss them a second time. In two steps, I reached the impressions in the clay and stepped firmly on them. I tried to look casual, even bored, despite the fact that my entire body was shaking. I felt tears well in my eyes.

"Nada," he said, panting and wheezing. "Puta madre!"

"I think you are getting old," I joked in Spanish. "There are no jaguar tracks here. Are you sure it wasn't a bird you shot at?" The comment had the desired effect: he laughed and took no suspicion from my face, which I tried to keep hidden—had he seen it, he would have known something was being kept from him. Watching his casual dimness, hearing his giggle, sent lava through my veins. In his face I saw the craven rapacity of my own species, the gluttonous waste; I saw the faces of elephants axed off while they screamed, bears strapped with bile extractors coming from their stomachs, a rhino bleeding from where its horn had once been. For a moment my hand tightened on the handle of my machete, and I thought of killing him myself, returning what he deserved. But my hand hung despondently at my side. I felt too hollow to do anything at all.

I was no longer in control of my emotions. As we made our way down the cliff toward the two boats below, I tried to get hold of myself, but I had been utterly disarmed by seeing the suffering cat. Tears welled in my eyes as I climbed back down the slope. Too defeated to care what happened next, I sold out and boarded their boat. Amid the carcasses and flies, I accepted a free ride and speedy, safe return to civilization. My great solo expedition was over.

18

The River

Ah how shameless—the way these mortals blame the gods.
From us alone, they say, come all their miseries, yes, but they
themselves with their own reckless ways, compound their pains
beyond their proper share.

—HOMER, THE ODYSSEY

If you journeyed back 2.7 billion years, just before photosynthesizing life began eating light and oxygenating the world, you would not be able to breathe. Plants literally created the livable reality that enabled air-breathing life-forms to exist.

In his book *The Sacred Balance*, David Suzuki takes a step further to point out that even if a time-traveling human came prepared with a breathing apparatus, life on pre-plant earth would be far from safe. Water would be toxic to drink because there would be no "plant roots, soil fungi or other microorganisms to filter out heavy metals and other potentially dangerous leachates from rock." There would also be nothing to eat. Even if a time traveler had anticipated the barren pre-life environment and brought seeds to plant fresh vegetables, he or she would find no soil to plant in "because soil is created when living organisms die and their carcasses mix with the matrix of clay, sand, and gravel." Looking backward at the earth in this way allows a sharper appreciation of the modern world, where plants and animals make up the living biosphere, in which life creates and sustains life. A walk through the woods starts to take on a completely different meaning.

By now, of course, plants have run the show long enough to put the more mobile animal species to work for them. Colorful flowers attract bees, birds, bats, and myriad other transporters of pollen; food in the form of ripe fruit is advertised in gaudy colors that are irresistible to the animals that will consume it and later deposit the seeds from it in their droppings. Many plants have evolved to depend entirely on animal dispersal.

Among the Amazon's largest and most impressive tree species is the Brazil nut tree, a towering 150-foot giant of the forest. Yet this massive pillar of the jungle depends on some of the smallest creatures to exist, such as species of orchid bees that enter their yellow flowers and transport pollen between male and female trees. Fertilized by the bees, the trees produce large, woody seed-filled pods, like super-thick coconuts, that grow in the canopy and harden before plummeting to the jungle floor like cannonballs, with the speed and weight to kill.

The pods are too hard for virtually any animal in the forest to penetrate and would remain trapped and rotting on the jungle floor if not for one creature: the agouti. Agoutis are a sharp-toothed rodent, roughly the size of a rabbit, that specialize in opening the hard pods, exposing the Brazil nut tree's nuts. Agoutis bury the nuts in the ground for later, and inevitably forget many of their caches.

Orchids, bees, trees, and agoutis—and the list goes on to include a species of arrow frog that guards its tadpoles exclusively in the empty Brazil nut pods that become filled with rainwater, and still others like the mosses, lichens, vines, bromeliads, ants, and other organisms that call the trees home. The result is a giant tree, a crucial element of life in the Amazonian landscape. Not to mention an important cash crop for the people who call the jungle home. Interestingly, Brazil nuts are one of the few products that come from the Amazon that can actually bene-fit the forest. Because of their complex life cycle, the trees are found only in primary forest, and so for Brazil nut farmers it is wiser to leave the forest standing and extract the Brazil nuts each year than to level the forest and attempt to farm in the poor soil. In this way the Brazil nut tree's clever lifestyle in the forest and resulting bargain with human industry has protected immeasurably vast areas of the Amazon.

In the wake of my solo trip I gradually made my way back to civilization. I spent more than a week camping and explor-ing deep in the back areas of Santiago's land, through forest riddled with pathways that led from one Brazil nut tree to an-other. I fished in streams, observed animals, and tried to take stock.

For some time the feeling of failure persisted. The plan had been to spend a week hiking upstream past the Western Gate, and then another week to ten days filming, and then return. In all I had budgeted just over three weeks for an expedition that in the end lasted less than two. Of the time I had spent alone I had filmed virtually nothing, saving battery and time for when I could set up, stake out, and concentrate on filming and not hiking. But the stakeout period never came. I would not, it seemed, be making a documentary after all.

Though I had cursed the bad luck of encountering the tribe, gradually my wonder replaced fear and disappointment. What they must have thought to see a white man standing at the edge of the forest? They had almost certainly never seen one before. Of course, many questions were left unanswered: what they thought, what they felt, and what would have happened if there had been less distance between us. In hindsight, I wish that I had waved to them, that there had been some form of communication other than the tense staring back and forth. But it was the jaguar's eyes that continued to haunt me, and the question of whether she had lived consumed my thoughts day and night.

Beyond the immediacy of her pain and needless death, the jaguar encounter was significant as part of the larger principle of the human relationship with predators. Only recently, scientists have begun to realize the true impact of apex predators on ecosystems; the parable of wolves in Yellowstone National Park has become the classic example. The last wolves in the Yellowstone area were killed in 1926, and the park remained without them until scientists reintroduced the species in 1995. The resulting changes that the predators caused in the ecosystem shocked everyone. For decades, the elk herds had overpopulated and overgrazed the riversides, munching the aspen, willow, and other tree saplings before they could mature. With the reintroduction of wolves, the elk numbers fell by half, and their behavior changed—they move more often now, because even when the wolves aren't around, the elk are on guard. With the elk moving often and in fewer numbers, new trees are able to grow, for the first time in almost a century. The rich new foliage allowed beaver numbers to increase, which in turn had positive impacts on the fish population. Coyote numbers also fell sharply under the rule of wolves, allowing more rodents, rabbits, and small mammal life to flourish—this, combined with the increased fish stocks, benefited raptors like the bald eagle. With wolves culling coyotes, there are more red foxes; with willow trees growing, there is a greater diversity and abundance of songbirds.

The reintroduction of wolves snapped a malfunctioning system rapidly back into order. Because the effects were so clear, it has become the textbook example of how apex predators improve the environment. From a human perspective, the benefits of top predators and the fully functional ecosystems they promote include healthier fish and game for us, healthier forests and timber, cleaner water, flood control (due to the flood-absorbing power of forests); there are myriad others. The scope and nuances of these top-predator influences, from wolves in Yellowstone to tigers in Asia, to sharks in oceans all over the world, are still little understood. What we do know is that these "umbrella" species are responsible for the health and balance of the systems they rule. By extension, it is not difficult to imagine the far-reaching repercussions that would take place in the Amazon with the loss of any of the top players: anacondas, black caiman, giant river otters, and jaguars.

Gowri arrived ten days after the solo expedition ended, after finishing a semester in New Jersey. Once I had crushed her in my arms for almost an hour, we spent a day in Puerto Maldonado together visiting cafés, catching up, and eating ice cream. I told her all that had happened. In the following days, she, JJ, Pico, and I, along with the other brothers, began preparing for another volunteer group. Months earlier I had coordinated with people from the United Kingdom, France, the United States, and even with a couple from Finland.

The first two weeks of the trip were spent at Saona, in Infierno. But for this group we had planned to spend some days in true Amazon adventure fashion: a boat expedition. JJ and

I agreed there was nowhere more perfect than Las Piedras. After experiencing the front-country feel of the Saona station for weeks, the volunteers were awed by Las Piedras's unbroken primary forest. JJ and I knew the guy whom the foundation had hired to watch Las Piedras Station, and he'd agreed to allow us an afternoon to walk the trails, without the foundation's knowledge.

Guiding in the rainforest is never easy. Your task is to show people wildlife that for the most part does not want to be seen and has spent millions of years adopting habits and camouflage that make it virtually undetectable. It doesn't help that most people assume that they can stomp through the forest with the noisy heel-toe city shuffle, while talking, and still see animals. But this group was different. They were careful, quiet, and tuned in. Two had come specifically to see anacondas, and others had significant outdoor experience. As a result, on the first night in Infierno I was able to bring them within ten feet of a mother tapir and her baby, near a colpa. On transects we'd seen monkeys, tamandua, and a huge arboreal porcupine. After spending a quiet night in a hammock, one of the guys in the group had returned to tell me he'd seen a jaguar with no spots, which I explained was a puma—a very rare sighting. With her skilled eyes, Gowri added greatly to what the group was able to see by picking out dozens of the smaller wonders of the forest that even JJ and I often missed.

This group enjoyed the forest wonderland and bare-bones station at Infierno in a way few ever have; they really got it, and as a result were able to have unique encounters with wildlife. But when we arrived at the Las Piedras Biodiversity Station, I saw awe on their faces. Their eyes were pulled upward by the pillars of huge hardwood trees that dominated the ancient forest, and before we even reached the station they remarked on how different and wise the forest seemed. When we reached the clearing and the actual station, with its palm-thatched roof, inviting hammocks, and paradisiacal array of flowers against the greenery of the clearing, there was a collective hushed silence while each person took in the beauty in his or her own way. More than once, people quietly exclaimed that they had never seen anyplace so beautiful.

While JJ told them the story of the station, and how we'd lost it, I slinked away into the kitchen to spend a moment with my own emotions and the flood of warm familiarity and longing that the station had cultivated. There was no denying it: Las Piedras was where I wanted to be; it was where I belonged. In those moments I savored the surroundings but inwardly cringed from the pain of renewed loss at seeing the old place again. Out on the deck, where memories of Lulu and other early adventures loomed like ghosts, I stood before the map of the Madre de Dios, tracing over the massive protected areas.

In the northwest was Manu National Park, where Charles Munn had begun his research before heading to the southwest of the Madre de Dios to focus on the Tambopata/Candamo project and create Bahuaja-Sonene National Park. To the northeast of the map was Alto Purus National Park, that mysterious roadless giant, home to nomadic tribes and the headwaters of Las Piedras. As I gazed at the map it became clear that in the center of these huge protected areas lay the majority of the Las Piedras River—a vast unprotected area with the power to connect the established national parks.

Munn had come to the Las Piedras River. He'd even built a lodge, JJ told me, with a Peruvian partner, a few hours from our station. But then he'd vanished from the Madre de Dios for reasons no one seems to know. Looking at the map, I wondered if he had secretly been planning a fourth giant park—it certainly made sense. The other three created a lopsided triangle, with the Piedras in the center; protecting it would connect the others and possibly make the Madre de Dios home to the largest area of protected west Amazon in existence. It would mean that almost the entire Madre de Dios was protected, ensuring survival for the ecosystem and species, as well as local people and the numerous isolated Indian groups.

In the months that followed, keeping Las Piedras off my mind was impossible. I spent hours dreaming of ways I could one day buy it, and dreaming of what it would be like to get it back. I would build a long network of canopy walkways at the station, and we'd have people year-round studying the wildlife, running the farm, and educating people from all over the world. We would have a rehabilitation center for wildlife, and maybe even specialize in giant anteaters. But most important, we could start truly working to get the entire river protected, and connect the national parks. Anything was possible.

My room at home in New Jersey was littered with notes and plans, funding proposals, and various other possible leads on the subject of reclaiming Las Piedras Station. I knew that the foundation that owned it was not using it and had no idea what to do with it. The preserve was surely a financial drain for them. And so after so much worry and want, planning and hope, I was completely blindsided to receive a phone call from JJ that started with the simple, ecstatic sentence: "We got it back!"

The return to Las Piedras was surreal. Apparently the foundation had purchased much of the reserve Emma and JJ had owned, but never paid up for the portion where the station was located. After a few years of stalling, not knowing what to do with the land, the foundation decided they wanted out and offered to just give it back. However, of the original 27,000 acres that Emma and JJ had protected, they gave back only about 1,200—just the part with the station on it. The rest of the land, the vast area of primary forest, they kept and planned to sell for almost five times the price they paid. Though the fate of the land remained uncertain, it was impossible to feel anything but wondrous gratitude at the news that the station was ours once again. The Las Piedras River would be home once again.

By that time, the logging road that had come off the trans-Amazonian highway years ago was well established. On our return to Las Piedras, JJ and I witnessed dozens of trucks per day. They were cutting the heart out of the river in a land that was too big for anyone to police. How humans love plundering a forest, like spoiled children with their parents' ATM card and no concept of moderation. Each truck that passed us, though, was fuel for the rekindled mission of getting the station running, getting universities and tourists to come, and making it known what a treasure the Las Piedras River really is. But before we could begin implementing our plans, we needed to address the fact that the station was in dire need of repair.

When the foundation left they took the dishes, couch cushions, our radio, and virtually everything else. They also left the roof in tatters, pocked with holes that let in rain, which threatened to saturate and destroy the wood of the deck. When JJ and I first returned, we faced months of worry as the rainy season approached, but in the end we were saved by dozens of past volunteers, friends, family, and people from all over the world who pitched in funds to fix the roof. Later, volunteers, friends, Gowri, JJ, my sister Michelle, and I hauled the materials up through the jungle. The support we received for the roof emergency was just one of the signs of how much Tamandua Expeditions had grown over the years.

In the waning days of 2012, I spent weeks at the station, walking trails and visiting my old friends, the kapok trees, and other denizens of the ancient forest. The peccary herds were still nowhere to be seen, but the jaguar tracks were as frequent as ever. I spent mornings watching the macaws bicker near the river and at night explored the swamps where the frogs were preparing for the start of the rainy season. I set up camera traps at the colpa deep in the forest, and weeks later was stunned by what they recorded.

On the main deck of the station, Gowri, JJ, and I watched in astonishment as everything from Spix's guans to tapir passed before the lens. There were curassows and pumas, ocelots, giant armadillos, howler monkeys, white capuchins, peccaries, and numerous jaguars. We gasped together at a video taken on Christmas Day when I had gone to check the cameras; it showed a tremendous male jaguar walking where I had knelt just moments before—clearly interested in my scent. All told, the two cameras recorded more than two thousand videos in four weeks, of every conceivable species in just a few square yards. Later on the videos would be combined as a five-minute short film about the colpa, which I titled *An Unseen World*. The short was covered by Mongabay.com and Yale Environment360, and was selected as a winner of the 2013 United Nations Forum on Forests short film contest. The diversity and abundance hidden in the forests of Las Piedras caught the imagination of people all over the world.

Yet, for me, the greatest discovery came from a camera that

was placed just behind the station itself. The short clip showed a massive mother anteater striding through the bright morning forest and approaching the camera. She sniffs, touching her long, black-tipped nose to the lens, perhaps recognizing human scent, before moving on. As she passes out of the frame there is a baby anteater visible, clinging to the wiry fur of her back. The clip stole my breath when I saw it. It is impossible not to wonder if the anteater in the video was Lulu, all grown up, with a baby of her own.

At a United Nations screening in Istanbul in April 2013 I watched as representatives from all over the globe narrowed their eyes in wonder at a place so wild. There was Lulu, the trees, the jaguars, and other Las Piedras wildlife. I thought with hope of Munn and the Candamo story. As I took the podium to address the audience, it was with Lulu, and the tribe, and the eyes of the wounded jaguar. I spoke of the major national parks, the triangle with Las Piedras at their center, the heart of the Madre de Dios, and for the trees and animals that cannot advocate for themselves.

In the end, it wasn't all about risking my life at the limits of the long headwaters beyond the Western Gate. In the end, the answer came where all the answers have come over the years, from the magic of the Las Piedras River, with its long, winding path, swift current, and seemingly endless expanses of riotous rich green canopy. As I look back, it seems that although I was born in Brooklyn, my life started on that river. It changed everything.

In the short time since we got the station back, many things have happened. Elías, JJ's nine-fingered brother, was elected president of Infierno. Pico recently received financial support from family and friends and flew first-class to Lima for surgery that straightened his legs and will allow him to walk more normally. In the preoperative psychological exam, the doctor told Pico that he was one of the most hyperactive patients he had ever seen—much to the pride of Pico. The Saona lodge in Infierno is continuing to grow, and the brothers are working out the chain of command as the wildlife and forest gradually regenerate. Everyone in the family still remembers that the forest and all it gives them are still there due to the wisdom and foresight of Santiago.

Gowri and I got married Hindu-style under the banyan tree we met beneath in India, with all our friends looking on and monkeys in the branches above. Months later, everyone joined for a church wedding in Manhattan. Yet during the ceremony, when the priest pronounced us husband and wife, I heard muffled laughs from friends in the pews who knew that Gowri and I weren't the only ones entering the union—that in my shoulder beneath the tux were two squirming botfly larvae. Each was as thick as a pencil and had refused to be removed after the most recent expedition. *I now pronounce you husband, wife, and parasites.* Part of Noel's best-man duties involved tugging out one of the worms the morning after the wedding.

Not long ago, while with a volunteer group, I met another giant anaconda at the floating forest. Her girth was slightly larger than a car's steering wheel, and this time I was with a crew of eight people, some bigger and stronger than me, in broad daylight. Dragging seven people on its back, the snake thrashed and curled, and pulled us down into the bog until we were up to our necks in water and there was no choice but to suffocate beneath the water as we sank, or release her. There are, it seems, some secrets the jungle is still unwilling to relinquish. But that won't stop me from looking. JJ, Pico, Gowri, and I are currently plotting our second expedition to a tributary that Don Santiago swore is home to a species of anaconda that has horns on its head. If it weren't for all that I've seen, I would never believe it, but as experience has shown, the man was never wrong.

With limitless expeditions ahead, and thousands upon thousands of miles to explore in the west Amazon alone, my childhood fears of being born in an adventureless age, at the end of all things, have long been assuaged. Many of the species I worried about as a child, ones that really were on the brink of oblivion, have been making comebacks: gorillas, right whales, bald eagles, polar bears, even amur tigers. The list is long. The success of these species and their respective ecosystems is a direct result of people fighting to save them.

In the Amazon the future is still very much uncertain, yet there are indications of change. The Madre de Dios recently trembled with the chop of propellers from teams of helicopter commandos sent by the government to remove illegal gold miners. The soldiers descended ropes and removed each miner from his forest-destroying gold barge, before blowing the barges apart so that they could never again ruin the land. Perhaps even more heartening, in 2011, years of indigenous protests finally resulted in the cancellation of plans to build a dam that would have flooded just under 114,000 acres of the Madre de Dios forests. Bolivians recently elected their first indigenous president, Evo Morales, and gave constitutional rights to the earth. Rivers, fish, air, trees—these things have rights there and are regarded as part of the collective public interest, of the inheritance that everyone is entitled to. Brazilian grocery stores were recently in the news for their decision to boycott Amazon-grown beef—one of the biggest drivers of deforestation. Hopeful signs, from the grassroots to corporate, governmental, and international levels across Amazonia, could fill volumes. It's not clear what will come of it all, and the chance of ruin is still very real, but there is light on the horizon. Yet there will always be challenges.

Currently a new, sinister threat is facing the Las Piedras, and in turn the Madre de Dios. This time in one of the most pristine areas of all, a place as remote and fantastically wild as the Western Gate: Alto Purus National Park. It is in these inaccessible rangelands that the Las Piedras is born. Geographically it is one of the largest swaths of intact old-growth, lowland rainforest left on earth. The uppermost Las Piedras, and the surrounding Purus region, cradles a virtually untouched wilderness where the pristine ecosystem is sanctuary to Amazonian wildlife that lives completely without human interference for hundreds of miles at a time. It is also one of the last strongholds for voluntarily isolated nomadic tribes, that have found refuge in this last, vast stretch of jungle—which may very well be about to have a highway cut through its heart.

While biologists, anthropologists, and numerous local communities within the park are aghast at the environmental and anthropological cost of the road, there are those who claim it is necessary to connect the few remote villages there, that are currently only accessible by air, to the rest of the country by road. It is a battle playing out in a corner of the earth so far removed from people's consciousness that the voices of the conservationists and indigenous communities who categorically oppose the road's construction are merely a whisper against the roar of those who crave development. Are we comfortable with knowingly extinguishing one of the last truly wild and authentic places on our planet? Are the ecosystem services, endemic wildlife, and already-imperiled tribal cultures there not worth fighting for?

The Purus road is a case of a few people interested in personal gain pushing for something that will allow them to profit more easily from exploiting the untouched forests there. And in pursuing their gain they will impoverish the world. Perhaps a few of them will get rich driving the last illegally cut mahogany trees out on the new road, or perhaps some of the hundreds of farmers that flock in will collect enough Brazil nuts to live well for some years. What is certain is that once the road is cut, settlers will invade. Vast tracks of forest will be leveled and burned for farms, and what was once pristine nature will be stamped out as they destroy one of the last remaining truly great wilderness areas. The wildlife will vanish, the tribes will retreat. Species will disappear forever. It is something being demanded by a select few, the results of which will forever erase a treasure from this earth. Our action or inaction will ultimately decide the fate of the mysterious unseen world that exists behind the veil of foliage and mist that is the Alto Purus watershed.

It is the paradox of Amazonian degradation that no one act will fell the giant, but instead a slow death by a thousand cuts— road by road, dam by dam, tree by tree; tiny slashes taken until the once-great tapestry is reduced to a ragged net. It is for this reason that the Purus road, a gash graver than most, cannot happen. It is for this reason that anyone who calls the jungle home, or has walked in the hallowed depths of the Amazon's shadows, or has even taken comfort in knowing that such places exist, is standing defiantly against the looming road. Such is our historic responsibility to the ecosystem and to wildlife, to the nomadic tribal cultures there, and to generations to come that will undoubtedly find it even more difficult to locate the visceral solace of untrammeled, pristine wilderness on earth.

The fact is that we are at a crossroads in history from which there will be no going back. The survival of the Amazon—or the rest of earth's rainforests and species, for that matter—is a question that will be decided in the next century. Future generations might only have film and fossils from which to learn about the great green wildernesses that used to exist. To them, the creatures that we know today could be as distant as mammoths and ground sloths are to us: nothing more than fascinating memories.

No. Passivity is a luxury that we cannot afford. Like it or not, this is our responsibility, and it will take every ounce of our will and ingenuity to meet it. After all, we are the first generation to be confronted with such a globally consequential task, against such overwhelming odds. From the Amazon, to the Congo, Sumatra, and Borneo—rainforests are vanishing day and night. The clock is ticking. And the only thing between the bulldozers and the trees is us.

It is curious that the Amazon, so large in our minds, is actually quite small. From far above on a clear day—the God's-eye view from a plane—the jungle looks like moss across an endless field. Even the clouds are far below, barely over the canopy, like bits of cotton blown across some lawn—or like mothers doting over the greenery, carefully watering the jungle from above. From so high up, the great jungle can be seen for what it really is: a very delicate collection of life on the smooth face of a planet. The greatest of its creatures become microbes, the great trees are reduced to lichen—even the bipedal primates that are so clever seem small and insignificant, save for the great scars they cut into the green. Cleverness is evident in this grand exploitation, but it will take a different kind of intelligence to preserve the whole and heal the wounds. And so the question becomes whether this one species can rise from cleverness to actual intelligence; to evolve from being a blight on the great basin to become instead the greatest of its stewards. But that remains to be seen.

Thankfully there are still vast stretches where the green is unbroken and the canopy extends to the horizon. On the Las Piedras River, beneath the towering treetops of bearded branches, dripping with oropendola nests and cascading floral tapestries, the jungle's morning chorus is as joyous as ever. No, I was not born in the wrong century, or even the wrong decade. This is exactly where I was meant to be, rafting beneath the tall emerald canopy, pulled by the swift current, on the long, winding, wild heart of the Madre de Dios.

ACKNOWLEDGMENTS

I may have never turned my journals, filled with impressions, memories, and sketches, into a book without the encouragement of my friends and family. Most important, I thank my parents, Edward and Lenore, and two older cousins, Danielle and Michael, who told me I could. Without their support and patience, this book would not exist. Along with my parents, my sister, Michelle, helped edit and shape early drafts. Also, for their thoughtful input and encouragement, I thank Peter Quartuccio, Lorraine and Douglas Capozzalo. As well as the generous and careful eye of my professor and friend Mike Edelstein, and those of my family and friends.

Most people see the jungle as frightening, but I assure you, as a first-time writer the world of publishing seems far more intimidating. Personally, the process has been a wonderful adventure, which is entirely due to the good fortune of having worked with some truly incredible people. For her care, compassion, and support—and for essential, technical help at the proposal—I must thank Linda Carbone. I am awed by my agent, Lindsay Edgecombe, whose renaissance abilities, energy, and dedication to the success of this book at every stage were invaluable. I must thank Harper-Collins for believing in this book, especially my fellow outdoorsman and very talented editor, Michael Signorelli, whose care, skill, and understanding helped shape this book into what it's become.

If it were not for the visionary people who were protecting wildlife and ecosystems long before I was born, an adventure like the one I had would not have been possible. I have written about many of my heroes in this book but in particular must thank Charles Munn, Celestino Kalinowsky, Daniel Winitzky, and the people of WWF, the Wildlife Conservation Society, and Conservation International. Also, I thank the people of Peru for the work they have done in the Madre de Dios and abroad. I am indebted to Steve Irwin for his incredible enthusiasm, passion, and skill—which I carefully studied and applied to my own animal interactions, and which has saved my life many times over. To Alan Rabinowitz for making such an impact on this world and on the species I love. To Jeremy Hance and the people at Mongabay.com, who actually report on things that matter, and who have shown such generous support for my work, thank you. A great thank-you to Bill McKibben for the books that I have read and reread, and for going out on a limb for me at a critical hour, when this book was in its infancy. Thank you to Dr. Jane Goodall for the stories my mother read me as a child, and for your kind words of support during the journey of this book.

I must thank JJ and Pico, Santiago and the entire Durand family, for welcoming me and making my relationship with the jungle possible. Yet, above all, my gratitude goes to the Amazon and the creatures within the forest, for allowing me a brilliant dalliance with their world, for shaping my life, and for permitting me to walk in its most removed and shadowed depths.

To my wife, Gowri, for her patience and support, her incredible courage, for more than I could write.

BIBLIOGRAPHY

Davis, Wade. *One River: Explorations and Discoveries in the Amazon Rain Forest.* New York: Simon & Schuster, 1997.

Grann, David. *The Lost City of Z: A Tale of Deadly Obsession in the Amazon.* New York: Vintage 2010.

Hemming, John. *Tree of Rivers: The Story of the Amazon.* London: Thames & Hudson, 2009.

London, Mark, and Brian Kelly. *The Last Forest: The Amazon in the Age of Globalization.* New York: Random House, 2007.

MacQuarrie, Kim. *Peru's Amazonian Eden: Manu National Park and Biosphere Reserve.* Barcelona: Francis O. Patthey, 1998.

———. *Where the Andes Meet theAmazon: Peru & Bolivia's Bahuaja Sonene & Madidi National Parks.* [Spain]: Jordi Blassi, 2001.

McMichael, C. H., et al. "Sparse Pre-Columbian Human Habitation in Western Amazonia." *Science* 336, no. 6087 (June 15, 2012): 1429–31.

Mother Nature Network. "Fact or fiction? 7 Eco-Myths Debunked." http://www.mnn.com/green-tech/research-innovations/photos/fact-or-fiction-7-eco-myths-debunked/the-rain-forest-is-a-nat#.

Pearce, Fred. *Deep Jungle.* London: Transworld, 2005.

Reuters. "Argentine Zookeeper Dies After Anteater Attack." April 12, 2007. http://www.reuters.com/article/2007/04/12/us-argentina-anteater-idUSN1235848120070412.

Revkin, Andrew C. "Murder on the Resource Frontier." *New York Times,* March 17, 2008. http://dotearth.blogs.nytimes.com/2008/03/17/murder-on-the-resource-frontier/?_r=0.

Schulte-Herbruggen, Bjorn. Project Las Piedras. 2003. http://www.savemonkeys.org/publications/report02.pdf.

Stewart, Douglas Ian. *After the Trees: Living on the Transamazon Highway.* Austin: University of Texas Press, 1994.

AUTHOR'S NOTE

As I send in the final edits for *Mother of God*, I find it exhilarating to have finished a project that has dominated the past four years of my life. Yet as this book ends, it is becoming increasingly clear to me that the story of the Madre de Dios is only beginning.

In the Madre de Dios, ensuring that protected land remains so is a full-time job; and on Las Piedras, we are only just beginning to make real strides. While we fight to improve the research station, unite landowners, and create a community of river guardians, we are forever in need of support, funding, and friends. If you would like to learn more about our current progress or how you can help or join our efforts in the Madre de Dios, visit www.TamanduaJungle.com and click on JungleKeepers.

ABOUT THE AUTHOR

Paul Rosolie is a naturalist and explorer who runs Tamandua Expeditions, which uses tourism to support rainforest conservation. He has worked on conservation projects in tropical ecosystems around the world. His documentary, *An Unseen World*, won the short film contest at the 2013 United Nations Forum on Forests. *Mother of God* is his first book.

www.PaulRosolie.com